AI, Machine Learning, and Blockchain for Next-Generation Spectrum Management

AI, Machine Learning, and Blockchain for Next-Generation Spectrum Management presents a forward-looking exploration of how artificial intelligence, machine learning, and blockchain technologies are transforming spectrum management in the era of 5G and emerging 6G networks. Bridging engineering innovation with regulatory insight, it offers a comprehensive guide to dynamic, secure, and decentralized spectrum strategies. With real-world case studies, simulation-validated architectures, and actionable policy frameworks, the book serves researchers, engineers, and policymakers shaping the future of wireless communication infrastructure.

Key Features

- Uniquely synthesizes AI, ML, and blockchain for intelligent spectrum governance
- Covers real-world deployments including CBRS, smart grid systems, and IoT-based intelligent transportation networks
- Comprehensively addresses both technical architectures and regulatory frameworks, including 3GPP Release 18, ETSI, and Ofcom standards
- Globally integrates perspectives from academia, industry, and policy institutions.

AI, Machine Learning, and Blockchain for Next-Generation Spectrum Management

Edited by
Vinod Hanmant Patil, Amol Krishnat Kadam,
Rajesh S. Prasad, Suresh Shanmugasundaram, and
Vinod Pachghare

CRC Press
Taylor & Francis Group
Boca Raton London New York

CRC Press is an imprint of the
Taylor & Francis Group, an **informa** business

Designed cover image: Vinod Hanmant Patil

First edition published 2027
by CRC Press
2385 NW Executive Center Drive, Suite 320, Boca Raton FL 33431

and by CRC Press
4 Park Square, Milton Park, Abingdon, Oxon, OX14 4RN

CRC Press is an imprint of Taylor & Francis Group, LLC

ISBN: 9781041222903 (hbk)
ISBN: 9781041222910 (pbk)
ISBN: 9781003731597 (ebk)

DOI: 10.1201/9781003731597

Typeset in Minion
by Newgen Publishing UK

Contents

About the Editors

Vinod Patil holds a Ph.D. in Electronics Engineering with specialization in Wireless Communication. His core research areas include wireless communication, cognitive radio, spectrum management, and smart grid technologies. With a strong academic background and over 18 years of teaching and research experience, Dr. Patil has made significant contributions to the field through research, innovation, and academic leadership.

He has published more than 30 papers in reputed Scopus and Web of Science-indexed journals and international conferences and holds granted and published patents in areas such as antenna design, smart grid management, and IoT-based systems. Dr. Patil is a recognized Ph.D. guide and reviewer for leading journals including *IEEE Transactions* and Taylor & Francis publications. His research achievements include developing cognitive radio testbeds, proposing energy-efficient routing algorithms for WSNs, and advancing green communication models.

Amol Krishnat Kadam is a Ph.D. in Computer Engineering and is recognized for his expertise in machine learning and software engineering. With an extensive academic and research background, he has secured major research grants from UGC and AICTE, including a ₹12 lakh project on software testing laboratory design and a project on wireless network security.

He has authored 37 international journal papers, with 20 indexed in Scopus and 2 in Web of Science and has presented his work at numerous international conferences. His current research supervision includes six Ph.D. scholars working on topics such as software defect prediction, gesture recognition, and CNN-based agricultural analysis. He also holds several patents, including one granted and others published or filed, reflecting innovation in smart surveillance, energy monitoring, and robotics.

Rajesh S. Prasad holds a Ph.D. in Computer Science and Engineering. With an academic career spanning over 28 years, he has held prestigious positions including Principal, Dean, and Professor at various renowned institutions. His research areas include artificial intelligence, machine learning, deep learning, cybersecurity, and smart healthcare systems.

Dr. Prasad has authored over 130 research publications including 66 journal articles, 55 conference papers, 12 book chapters, and 4 edited books indexed in Scopus. He has guided 7 Ph.D. scholars and reviewed more than a dozen Ph.D. theses. His contributions include 3 granted patents, 5 published/filed patents, 13 copyrights, and multiple funded projects from IEEE and academic institutions. His international collaborations, such as with the University of Oxford in cancer prediction research, and his roles on editorial and advisory boards, demonstrate his leadership and continued impact on global research and education.

Suresh Shanmugasundaram is currently a lecturer at the University of Wales Trinity Saint David, UK. He holds a Ph.D. in Information and Communication Engineering and has over 20 years of experience in academia and research. His academic journey includes key roles in Indian and UK institutions, with expertise spanning Artificial Intelligence, Machine Learning, Deep Learning, Data Analytics, and Cloud Computing.

Dr. Shanmugasundaram has authored more than 70 research papers in reputed international journals and conferences, many indexed in Scopus and Web of Science. In addition to mentoring doctoral scholars and participating in funded research projects, he has served as a reviewer and editorial board member for several international journals, reflecting his commitment to advancing global research in intelligent computing systems.

Vinod Pachghare is Associate Professor in the Department of Computer Science and Engineering at COEP Technological University, Pune. He holds a Ph.D. in Computer Engineering and brings over two decades of academic experience in teaching, research, and mentoring. His core areas of expertise include cybersecurity, network security, cryptography, and machine learning, and he has played a significant role in shaping advanced curricula and guiding multiple postgraduate and doctoral students.

He is widely published, with numerous papers in reputed international journals and conferences. He is the author of acclaimed books on network security and computer architecture, adopted by various technical universities across India. Recognized for his academic contributions, he is actively involved in reviewing scholarly publications and serving on academic committees. His work bridges foundational theory with real-world applications, advancing both research and pedagogical innovations in computer science.

Contributors

Deevyankar Agarwal
University of Applied Sciences, Dubai, UAE

Zaid Alsalami
College of Technical Engineering, The Islamic University, Najaf, Iraq

Lakshmipraba Balaji
D. Y. Patil International University, Pune, India

Gayatri M. Bhandari
JSPM's Bhivarabai Sawant Institute of Technology and Research, Pune, India

Archana Rajesh Date
Hon. Shri Babanrao Pachpute Vicharadhara Trust's Faculty of Engineering, Kashti, Ahilyanagar, India

Shubhangi Deshmukh
COEP Technological University, Pune, India

Gulbakshee J. Dharmale
Dr. D. Y. Patil Institute of Technology, Savitribai Phule Pune University, Pune, India

Nupur Gaikwad
Fr. C. Rodrigues Institute of Technology, Navi Mumbai, India

Tanaya Ganguly
Amity School of Engineering and Technology, Amity University Jharkhand, Ranchi, India

Amitesh Jaiswal
Bloomberg LP, New York, USA

Swati Jaiswal
Deccan Education Society Pune University, Pune, India

Kavita Joshi
D. Y. Patil Institute of Engineering, Management & Research, Pune, India

Radhika Joshi
COEP Technological University, Pune, India

Swati Joshi
D. Y. Patil Institute of Engineering, Management & Research, Pune, India

Amol Kadam
Bharati Vidyapeeth (Deemed to be University) College of Engineering, Pune, India

Sudhir Kadam
Ajeenkya DY Patil Innovation University, Pune, India

Munmun Kakkar
D. Y. Patil Institute of Engineering, Management & Research, Pune, India

Bharati Khatawate
Fr. C. Rodrigues Institute of Technology, Navi Mumbai, India

Renu Kachhoria
Vishwakarma Institute of Technology
Pune, India

Komala Soares
Government Polytechnic, Panaji, Goa, India

Akhilesh Kumar
COEP Technological University, Pune, India

Saikat Mondal
Amity School of Engineering and Technology, Amity University Jharkhand, Ranchi, India

Chetan More
Bharati Vidyapeeth (Deemed to be University) College of Engineering, Pune, India

Jyoti Sunil More
Fr. C. Rodrigues Institute of Technology, Navi Mumbai, India

Sourav Mukherjee
Amity School of Engineering and Technology, Amity University Jharkhand, Ranchi, India

Anuradha Nigade
Bharati Vidyapeeth (Deemed to be University) College of Engineering, Pune, India

Mangal V. Patil
Bharati Vidyapeeth (Deemed to be University) College of Engineering, Pune, India

Shrishail Patil
JSPM's Bhivarabai Sawant Institute of Technology and Research, Pune, India

Vinod H. Patil
Bharati Vidyapeeth (Deemed to be University) College of Engineering, Pune, India

Manali Pednekar
D. Y. Patil Institute of Engineering, Management & Research, Pune, India

Tushar Phadtare
JSPM's Bhivarabai Sawant Institute of Technology and Research, Pune, India

Prachi Pol
COEP Technological University, Pune, India

Rajesh Prasad
Bharati Vidyapeeth (Deemed to be University) College of Engineering, Pune, India

Deepak Kumar Ray
Bharati Vidyapeeth (Deemed to be University) College of Engineering, Pune, India

Kakade Sandeep
Vilasrao Deshmukh Foundation Group of Institutions Latur, India

Deepshikha Sarjare
Fr. C. Rodrigues Institute of Technology, Navi Mumbai, India

Arundhati A. Shinde
Bharati Vidyapeeth (Deemed to be University) College of Engineering, Pune, India

Nitin Shivale
JSPM's Bhivarabai Sawant Institute of Technology and Research, Pune, India

Anurag Shrivastava
Saveetha School of Engineering, Saveetha Institute of Medical and Technical Sciences Chennai, Tamil Nadu, India

Vijay Sonawane
JSPM's Bhivarabai Sawant Institute of Technology and Research, Pune, India

Spandan Surdas
Khoury College of Computer Sciences, Northeastern University, Boston, USA

Asheesh Tripathi
Commonwealth Cyber Initiative, Virginia Tech, USA

Divyansh Washi
Amity School of Engineering and Technology, Amity University Jharkhand, Ranchi, India

Preface

One of the most important and limited natural resources that form the basis of the current wireless communication is the radio frequency spectrum. Mobile broadband and satellite services, the Internet of Things, and autonomous vehicle networks are just some examples of virtually all types of the connectivity that rely on effective and reliable access to the electromagnetic spectrum. Traditionally, spectrum has been handled by the means of fixed, centralized allocation systems, whereby regulatory bodies license frequency bands to licensed operators on long-term basis under fixed conditions. Although this method brought stability in the early radio communications, it can no longer be used to meet the requirement of the fifth-generation (5G) networks, and indeed the sixth-generation (6G) systems which are currently being developed in the research laboratories and international standards organizations.

The dynamically rising number of connected devices, data-sensitive applications, immersive multi-media service and machine-to-machine communication has taken the traditional spectrum governance to the brink of collapse. Investigations by the Federal Communications Commission and the International Telecommunication Union have repeatedly demonstrated that substantial amounts of licensed spectrum are left underutilized at any particular time and place and yet operators are reporting overload and capacity issues elsewhere. This paradox scarcity and waste at the same time is structural failure of the static version of allocation, and this has brought the creation of intelligent, adaptive, and trustful spectrum management to be one of the defining issues of the next decade of wireless engineering.

The capabilities to tackle this challenge can be provided by artificial intelligence, machine learning, and blockchain technology, which also have different but complementary capabilities. Machine learning and artificial intelligence can cause spectrum systems to perceive their radio environment, anticipate traffic patterns and spectrum availability, and optimally assign resources on the fly and identify and remediate interference automatically. Spectrum sharing between operators who are unfamiliar or distrustful of each other requires the decentralized, tamper-resistant governance infrastructure offered by blockchain technology that logs spectrum transactions cannot be altered by any party, implements sharing agreements based on smart contracts and offers audited records to ensure compliance with regulatory authorities. The combination of these technologies forms the basis of intelligent enough autonomous spectrum systems that are trustworthy enough to manage and scalable enough to serve the billions of devices that 6G networks will be required to serve.

This book is a compilation of these developments under one roof. The literature on AI and machine learning-based spectrum management, as well as on blockchain-based spectrum governance, is abundant and quickly expanding and yet still scattered across hundreds of journal articles and conference papers. The current efforts of researchers in deep reinforcement learning of dynamic spectrum access have little visibility of the blockchain governance efforts underway; the architects of federated learning are generally not aware of the regulatory and standardization efforts underway at ETSI, Ofcom and 3GPP Release 18. This book is presented as a synthesis – a source that has the underlying theory, the technical proceedings, the governance systems and the practical implementation all under one roof organized in a manner that aids researchers as well as practitioners.

The ten chapters within this volume are based on the work of about forty three contributors based in academic institutions, industry as well as independent research based in India, the United Kingdom and the United States. After a comprehensive editorial review, chapters were then shortlisted and grouped to provide thematic coherence, depth progression and consistency to the overall coverage scope of the book as stated. The outcome is an overview of a volume covering all sides of the next-generation spectrum management, including the historical and regulatory framework, the AI and machine learning approaches, the blockchain and data control models, down to the practical implementation in the smart transportation systems.

The book is aimed at graduate students and doctoral researchers in electronics engineering, computer science, and telecommunications; practising engineers in the design of 5G and 6G networks, spectrum planning, and IoT system deployment; spectrum regulators and policymakers aiming to operate frameworks of dynamic licensing and spectrum sharing; and cross-disciplinary scholars interested in spectrum governance through the lenses of economics, law, or information systems. It is not meant as an introductory text in the fields of wireless communications, machine learning or blockchain technology; the readers are expected to possess a basic knowledge on the principles of at least one of those fields.

ORGANIZATION OF THE BOOK

This volume consists of ten chapters which are arranged in four sections and each part is based on the other. Part I introduces the contextual information, the historical, regulatory, and architectural background on which all other chapters in the book rely. Part II is the artificial intelligence and machine learning techniques that constitute the technical core of next generation spectrum management. Part III touches on the blockchain, governance, and security layer that enable these AI-driven systems to be used in real and multi-operator, commercially complex networks. Part IV completes the volume with an application study based on reality that illustrates how all of these themes come together in a working system.

Chapter 1 follows the development of spectrum management to trace the history of spectrum management since the command-and-control period of the early twentieth century to the dynamic and AI-driven paradigms of 6G, which will inform all the other chapters. Chapter 2 offers the architectural reference concept of 5G and 6G systems including 5G core network architecture, spectrum band taxonomy, Massive MIMO, network slicing, intelligent reflecting surfaces, and integrated sensing and communication vision of 6G that others at AI and blockchain chapters follow.

Chapter 3 intersects between classical cognitive radio theory and modern AI and blockchain systems, inspired by direct testbed experience to realize why classical designs are not successful and what drives the hybrid systems that ensue. In Chapter 4, a detailed description of architectural and algorithmic solutions to AI driven spectrum management in 6G networks is given, including supervised, unsupervised, and reinforcement learning methods in a systematic, pedagogically friendly treatment. Chapter 5 introduces ground-breaking AI systems to predict and optimize dynamic spectrum with the latest transformer-based and hybrid deep learning systems of the research frontier of 2023–2025. Chapter 6 deals with spectrum forecasting in ultra-dense IoT settings, where most spectrum management studies, of reliable sensing, sufficient computation, and continuous connectivity, fail and federated and edge-assisted learning become the correct paradigm.

Chapter 7 gives extensive discussion of blockchain-enabled spectrum auctions and regulatory rules, the entire history of spectrum regulation, smart contract designs of registering bidders and determining winners, and machine learning algorithms of predicting auction prices, detecting fraud, and optimizing reserve prices. Chapter 8 introduces a full hybrid AI-blockchain system to govern spectrum data in smart grids in a way that is secure, fair, efficient, and reduces carbon, and evaluates the outcomes of privacy, equity, and efficiency and carbon reduction metrics. Chapter 9 places AI-blockchain spectrum systems in the larger context of world regulatory frameworks, issues of security and privacy, ESG models, and spectrum data governance architecture – the chapter introduces the concept of spectrum data lakes as base infrastructure to the next-generation spectrum management.

Chapter 10 ends the volume by reviewing and introducing an original case study of using blockchain to facilitate Intelligent Transportation Systems throughout the 5G-to-6G development. The field deployable prototype of the On-Board Unit in the chapter, which is tested on real urban roads with vane nodes of five vehicles and real-time ML-based traffic classification, draws the volume to the relevant, concrete conclusion to prove the practicality of all its major themes.

The editors are indebted to the authors who contributed in the volumes with their expertise, commitment, and generosity that are scholarly. The editorial and production staff of CRC Press and Taylor and Francis group has also been very helpful in helping us with the publication process. We believe that this book will become a valuable and lasting resource to the researchers, engineers, and policymakers as the sphere of the next-generation spectrum management keeps developing.

The Editors
Dr. Vinod Hanmant Patil
Dr. Amol Krishnat Kadam
Dr. Rajesh S. Prasad
Dr. Suresh Shanmugasundaram
Dr. Vinod Pachghare
March 2026

CHAPTER 1

Evolution of Spectrum Management for 5G and 6G Networks

Gulbakshee J. Dharmale, Sourav Mukherjee, Divyansh Washi, Tanaya Ganguly, and Saikat Monda

1.1 INTRODUCTION

The development of the wireless communication is dependent on the management of its basic and finite resource: the radio frequency spectrum. Every generational leap, from the analog voice calls of 1G to the high-speed mobile broadband of 4G, has been enabled by innovations on the spectrum offering, access, and use. Given the world's insatiable demand for data, fueled by the explosion of connected devices and data-rich applications, the fifth-generation (5G) and the upcoming sixth-generation (6G) networks need a radical change from the historical paradigms of spectrum management [1]. This paper chronicles the transformative evolution of spectrum management from the command-and-control rigid paradigms of the past to the dynamic intelligent and cognitive ones necessary for a hyper-connected future. For a long time now, spectrum management has been characterized by a static spectrum allocation model. After all, regulatory authorities set frequency blocks and issue long-term exclusive licenses for use to network operators. While this model worked in the past, it now leads to significant inefficiencies and wastes resource cartridges in the fast network of

DOI: 10.1201/9781003731597-1

today. This legacy model generates artificial scarcity of radio spectrum, resulting in underuse of valuable frequency bands. Licensed spectrum, in fact, is most underused [2]. The implementation of 5G networks has introduced new technologies which aim to increase flexibility and efficiency in how the spectrum is used. One of these technologies is dynamic spectrum sharing (DSS), which allows 4G LTE and 5G New Radio (NR) to operate in the same frequency band [3]. DSS gives telecom operators the ability to switch and customize the spectrum resources between 4G LTE and 5G on a millisecond basis. This technology makes it easier and less expensive for telecom operators to migrate to 5G. It allows operators to fully use the 4G spectrum and 4G LTE infrastructures in place. DSS works with other technologies such as software-defined networking (SDN) and network functions virtualization (NFV) to create new technologies. One of these new technologies is network slicing, which enables operators to divide a single physical network into several virtual, specialized networks [4]. Operators can customize the spectrum resources in the physical network to satisfy the unique quality of service (QoS) needs of each 5G use case. These use cases include enhanced mobile broadband (eMBB), ultra-reliable low-latency communications (URLLC), and massive machine-type communications (mMTC). 5G also expanded use of the technology into the mmWave bands. This advanced technology allows 5G to use new bandwidth. This new bandwidth introduces new advanced challenges such as path loss and blockage which also need advanced techniques to mitigate [5]. The construction of 5G systems demands new intelligent decision-making, moving away from the classical automated systems and rule-driven methods. This is where artificial intelligence (AI) and machine learning (ML) come into play [6, 7, 8]. With the ML applications, especially deep reinforcement learning, systems can automate, in real-time and remotely, the decision-making processes of controlling, predicting, managing, off-loading resources, and dynamically managing inter-system interferences [9, 10, 11]. Also, the systems with real-time adaptive and AI-assisted deep learning spectrum managers remove the spectrum recovery and reactively monitored spectrum and management controls restoring manual spectrum to off-load preventive, predictive, and self-optimizing spectrum management to AI and automated systems.

6G will finish the final segment of the networks evolutionary cycle and integrate intelligence into the core of the network's architecture transforming it into what is termed as an "AI-native" network [12]. 6G will operate on the far reaches of the terahertz (THz) spectrum (0.1–10 THz) and promises to deliver terabit-per-second data rates, albeit with far greater challenges on the physical layer [13,14, 15]. Most importantly, 6G will redefine the fundamental interconnections between the network and its surroundings. Intelligent reflecting surfaces (IRS) will revolutionize the wireless ecosystem by turning it from a passive medium into an active, programmable, software-defined zone that can focus and redirect signals on demand to improve coverage and capacity [16,17,18]. The integrated sensing and communication (ISAC) systems can additionally use a single waveform and hardware to support each of the various functions associated with radar sensing (radar) and multiple communication data streams (communication), and thereby achieve an extremely efficient use of the spectrum and allow for numerous new forms of environmental awareness and applications [19]. Autonomous, intent-based, cognitive spectrum orchestration through convergence will provide 6G networks with the capability to manage their resources with AI. The agents will be responsible for the intelligence and the efficiency. As a direct result of the transformational process and understanding of the evolution of the spectrum, the world will be waiting for these networks. Therefore, it will build the future of connected 6G networks; that is the goal of this paper.

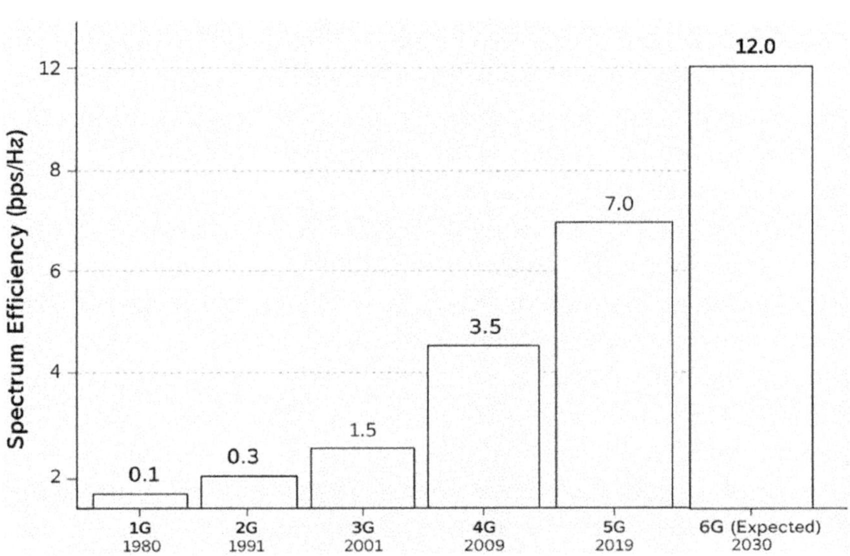

FIGURE 1.1 Spectrum utilization efficiency across generation (1G-6G). [36]

1.2 THE EVOLUTION OF SPECTRUM MANAGEMENT

The history of spectrum management has been characterized by dealing with an array of multifaceted and sophisticated approaches in the optimization and enhancement of management functions. Managing the spectrum's radio frequencies requires significant adjustments due to the evolution of wireless communication technologies reaching the mobile broadband and data service levels. The transformation of spectrum management systems from rigid and static allocations to flexible and dynamic frameworks has not occurred in a single stroke. It is the result of a gradual evolution in technologies and policies that prepare the world for the 5G and 6G generations [20].

1.2.1 The "Command-and-Control" Era (Legacy Models)

In the 20th century, for a long time, spectrum management was based on the "command-and-control," or fixed spectrum allocation (FSA) approach [2]. During this period, the spectrum was considered property and the U.S. FCC and the International Telecommunication Union (ITU) issued exclusive use licenses to broadcasters and mobile operators leasing blocks of frequencies for a long time. Spectrum management was primarily focused on the administrative mechanisms for order and service interference avoidance. The administrative simplicity of the FSA model and the associated "predictability" were key advantages. License holders received exclusive, interference-free use of their frequencies, and FSA approach "predictability" was a considerable incentive for the large national cellular and broadcasting infrastructures, which required large and sustained capital investments to establish. The "order" provided through the FSA model was essential for the relatively static demands of early digital services and analog voice communications.

Still, the most important problem with this model is its inefficiency with respect to spectrum. The command-and-control model granted exclusive rights to large portions of the spectrum, which created artificial scarcity. A large body of research demonstrates that much of the licensed spectrum is underutilized and even completely idle at certain times or certain places. This is what creates spectrum holes or white spaces—licensed frequency bands that are not being used. The most famous example is the TV white spaces—which are the idle channels that sit between active channels of broadcast television. The inability of the system to respond dynamically to changes in traffic load created huge inefficiencies, resulting in

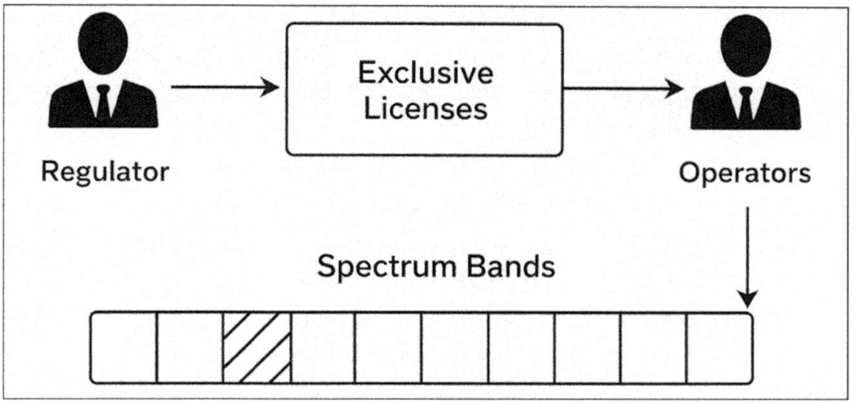

FIGURE 1.2 Fixed spectrum allocation (FSA) model illustrating exclusive licensing and underutilized spectrum bands. [2], [20]

unprecedented wastage of the spectrum resources and massive bottlenecks at the time when the demand for wireless data was rapidly increasing.

1.2.2 The Shift to Flexibility (Pre-5G Innovations)

The inefficiencies associated with FSA led to further focus on research and innovation on developing flexibility in spectrum use. The cognitive radio (CR) concept's initial idea was put forth by Joseph Mitola in 1999 [21]. A cognitive radio system is an intelligent wireless system that 'learns' its radio frequency environment and adapts its parameters like frequency, power, and modulation to avoid interference and allow efficient communication [22, 23].

The operational embodiment of the cognitive radio concept refers to dynamic spectrum access (DSA) [22, 23]. DSA frameworks work to maximizing the efficiency of spectrum utilization. Spectrum access is permitted to unlicensed, or "secondary," users to opportunistically access band-licensed spectrum whenever the "primary" licensed user is not fully utilizing it. This involves spectrum under-usage and opportunistic access permitted to unlicensed users. This captures the essence of the DSA concept in its simplest. This involves two basic functions:

Spectrum sensing – The cognitive radios use the sensing of the spectrum in order to determine if there are primary users present in the spectrum. Once determined that there are no primary users present, they are then able to identify vacant frequency bands (or "holes") within the spectrum. A variety of techniques can be used for determining when a vacant frequency band exists, such as energy detection, matched filter techniques, and the use of cyclostationary features to assist with the identification.

Opportunistic spectrum access (OSA) – Once a vacant opportunity has been identified, the secondary user will have temporary access to it. However, there is one caveat to this, the secondary user must vacate the channel immediately upon discovery of the primary user transmission. If the secondary user does not vacate the channel, there could be an interference issue created between the two.

While OSA demonstrated that dynamic access is possible, it's purely opportunistic nature provided almost no assurances for commercial services. It is these developments that have inspired the creation of more formalized models for spectrum sharing, such as licensed shared access (LSA) in Europe and Spectrum Access Systems (SAS) in the United States. The most well-known realization of this concept is the Citizens Broadband Radio Service (CBRS) in the U.S. 3.5GHz band [24]. CBRS establishes an advanced three-tier sharing architecture that is coordinated by a cloud-based SAS:

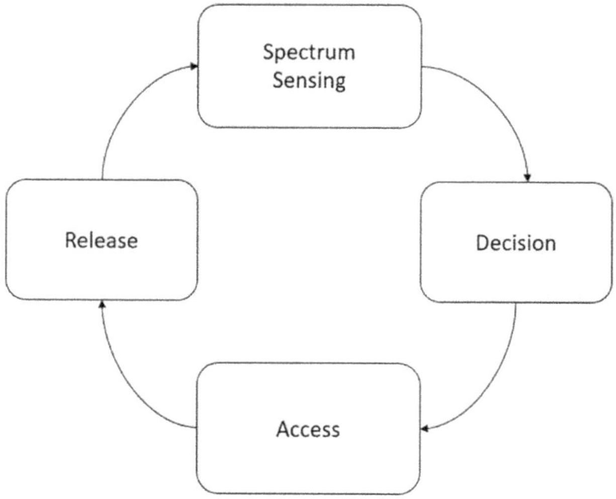

FIGURE 1.3 Dynamic spectrum access (DSA) and cognitive radio operation. [21–22]

I. Tier 1: Incumbent Access: The highest level of priority and protection is given to present federal users, especially to the U.S. Navy's shipborne radar systems. These users operate with absolute priority at all times.

II. Tier 2: Priority Access License (PAL): This tier is comprised of licensed users who through competitive bidding gain 10-year licenses for 10 MHz channels. PAL users are assured protection from interference from the tier below; however, they must not interfere with incumbent users [25].

III. Tier 3: General Authorized Access (GAA): This tier is available to all users on a licensed-by-rule basis and provides opportunistic, unlicensed-like access to any portion of the band that is not utilized by the higher priority tiers. GAA users have the lowest priority and must accept interference from Priority Access License (PAL) and Incumbent Access users.

Signing the CBRS framework shows you have succeeded in spectrum management through the creation of an efficient-sharing system for the coveted mid-band spectrum between commercial and government entities and the proliferation of private LTE and 5G networks.

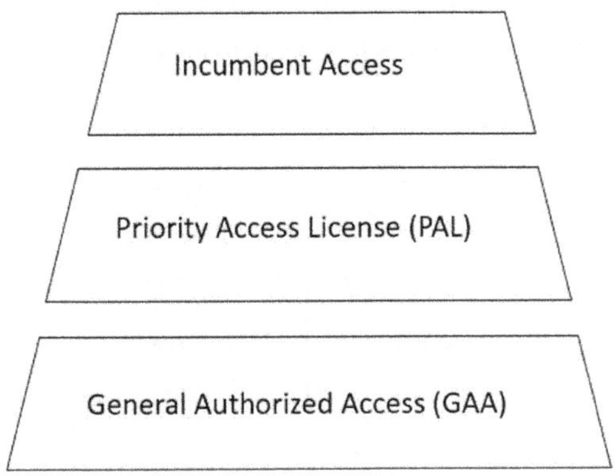

FIGURE 1.4 The CBRS three-tier sharing model showing the hierarchical relationship between incumbent access, priority access license (PAL), and general authorized access (GAA) users. [2], [20], [24]

1.2.3 The 4G/LTE-Advanced Contribution

While the shift toward more flexible use of spectrum is based on sharing technologies, the cellular standards themselves have also been modified to include new forms of dynamic spectrum use. Two of the primary factors in making 5G technology as flexible as it is being both part of the basic 4G LTE-Advanced standard. These are carrier aggregation (CA) and frequency band aggregation (FBA). Carrier aggregation allowed an individual mobile user equipment to simultaneously communicate with multiple frequencies at the same time (or "bands") which created a larger bandwidth than a single band, increased peak data rates and total network capacity when combining bands from different frequency ranges. While not specifically a method of dynamic spectrum access, it did allow for the integration of various spectrum resources to increase the maximum achievable rate of 4G networks to multi-gigabit per second levels.

Based upon this concept, 3GPP introduced licensed-assisted access as part of Release 13 [26]. The LAA is an alternative version of carrier aggregation that allows mobile service providers to utilize additional spectrum through combining their own licensed spectrum along with the unlicensed 5GHz band (which is the exact same spectrum that is being utilized by Wi-Fi), providing a significant increase in overall system capacity, particularly within high-density indoor environments such as stadiums or shopping malls. In order to ensure fair coexistence with Wi-Fi or other unlicensed users, LAA has included a "Listen-Before-Talk" (LBT) policy. The LBT policy requires the base station to listen to the channel for a couple of seconds prior to sending data. LAA is the first formalized standards-based integration of licensed and unlicensed spectrum and represents a fundamental principle of hybrid access, which is an essential component of the 5G Spectrum Strategy.

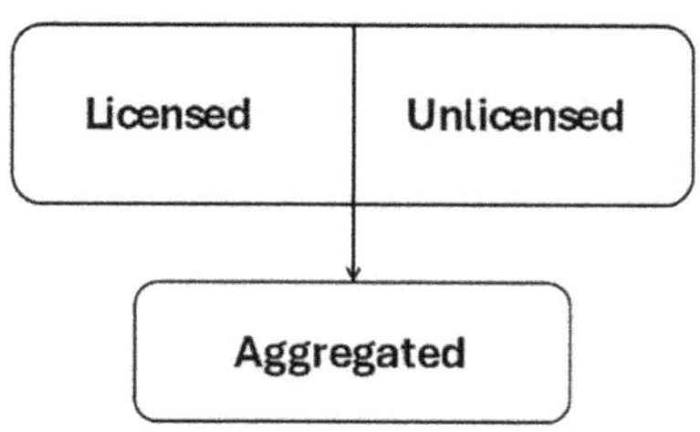

FIGURE 1.5 Carrier aggregation and licensed-assisted access (LAA). [2], [7], [25]

1.2.4 The 5G Conundrum: A New Dimension of Challenges

In terms of both flexibility and dynamic nature, the paradigmatic shift brought about by 5G creates an entirely new and very functional, multiple-faceted and complex set of challenges for spectrum management. The goal of 5G is not just to give us a lot faster mobile broadband, which is what previous generations of mobile technology gave us at a pretty uniform level. 5G will be giving us broadband for a wide range of different applications. In addition to addressing the traditional spectrum management (Triad Spectrum Management Strategy), the new 5G spectrum technologies present a number of issues that the manager needs to consider including: (1) Conflicting and Overlapping Varying 5G Services; (2) Exploding Network Interface with Network Densification and (3) Old Technology vs New Technology Spectrum Sharing Issues. Each of these issues have some type of interaction with each other as well. This creates a situation where the spectrum management challenge presented to it cannot be solved using its legacy management approaches.

The first wireless commander, Generation 5G, will have a three-tiered structure low, mid, and high that are physically very different from one another, and therefore functionally as well, and will be used to enable nearly all of 5G's objectives. Therefore, using multiple bands i.e. to provide both high capacity and multi-Gigabit speeds and ubiquitous coverage will present management challenges that are difficult to manage individually across each band.

In addition to providing increased variety in resources, 5G represents a significant change in the paradigm for managing those resources. The flexibility provided by the next generation of wireless networks is enabled through Network Slicing [27]. The major spectrum challenge of network slicing arises from the unbalanced and conflicting demands of different slices.

To meet its capacity and speed targets in the mid- and high-frequency bands, 5G requires considerable network densification. Instead of

TABLE 1.1 Comparison of 5G Service Categories [8] [36]

	eMBB	**URLLC**	**mMTC**
Throughput	High	Low	Moderate
Latency	Relatively high	Very low	Relatively high
Reliability	Moderate	Very high	Moderate
Connection Density	Low	High	Very high

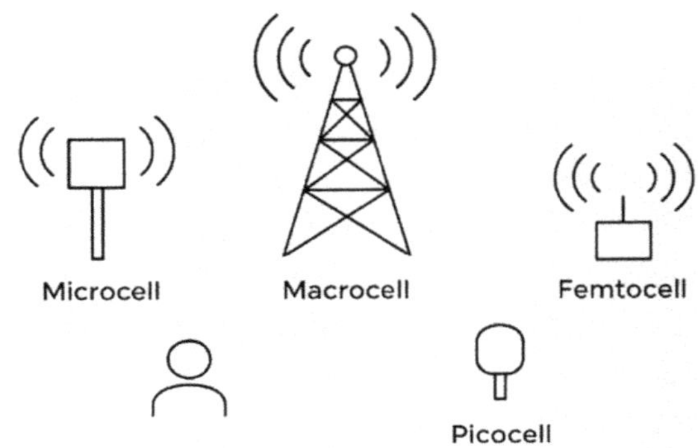

FIGURE 1.6 Heterogeneous network (HetNet) architecture. [7], [26]

employing just the traditional large macro-cells, the network architecture now includes Heterogeneous Networks (HetNets) where the macro layer is supplemented with a dense network of low-power small cells like micro-cells, pico-cells, and femto-cells. Small cells facilitate greater data rates and improved spectral efficiency, as the network infrastructure is physically brought closer to the user through aggressive frequency reuse. But network densification brings a completely new challenge: interference [27]. Each small cell in a dense HetNet grid may become a source of interference for surrounding cells, and this increases the potential for destructive co-channel interference and adjacent-channel interference.

1.2.5 Dynamic Spectrum Sharing (DSS)

To accelerate the deployment of 5G, particularly in the low and mid bands already saturated with LTE, carriers have begun the use of DSS [3]. This technique allows time and frequency resources to be shared, and reallocated in real time to service LTE and 5G NR users in the same frequency band. Rather than going through expensive and time-consuming "hard refarming" to shift 5G to lower bands, operators can use DSS to "soft refarm" LTE bands, sharing the spectrum down to the millisecond. As DSS is low cost, operators can use it to claim 5G wide area coverage and enhanced connectivity, satisfying users migrating to 5G while upgrading existing spectrum and radio hardware through software.

DSS still has its challenges. The foremost of which is the substantial overhead involved in the management of two different radio access technologies operating in the same band. Both LTE and NR use control/reference signals that must be transmitted using their respective resource allocations before

TABLE 1.2 Classification of Primary Challenge Domains with Corresponding Specific Challenges and Causative Factors

Primary Challenge Domain	Specific Challenge	Key Characteristics and Causative Factors
Multi-Band Management: A Band for Every Use Case	Low-Band (Coverage Layer- Less than 1 GHz) Challenge: Spectrum Scarcity andReframing	Extremely valuable and contested asset. - Requires migration of legacy services (2G/3G, digital TV). - Involves a political, technical, and economic balancing act. [8][36]
	Mid-Band ("Goldilocks" Band- 1 to 6 GHz) Challenge: Incumbent Service Interference	High risk of blocking weaker satellite (FSS) signals. - Requires mitigation via Protection Zones, advanced filtering, and guard bands. [8]
	High-Band (Capacity Layer: mmWave>24GHz) Challenge: **1.** Physics: Poor Propagation: the biggest challenge to mmWave technology. In comparison to lower frequency bands, mmWave frequencies have an extreme amount of path loss and are much more susceptible to being blocked by walls, leaves, windows, and even a users' hand. **2.** System Complexity: As mmWave systems suffer from large amounts of signal loss during propagation, they employ more advanced forms of antennas as part of their technology, especially beamforming using massive MIMO (multiple input multiple output) arrays, which are formed by hundreds of individual antennas.	Physics (Poor Propagation): High path loss and susceptibility to blockage (walls, foliage, rain). - Limits coverage to short ranges (a few hundred feet). - Results in unreliable links dependent on environment. [5] System Complexity: Reliance on Massive MIMO and beamforming for pencil beams. - Requires significant processing power for beam management and user tracking. - Increases cost and complicates mobility support.

(*continued*)

TABLE 1.2 (Continued)

Primary Challenge Domain	Specific Challenge	Key Characteristics and Causative Factors
Service Diversity & Resource Allocation eMBB, URLLC, mMTC)	Conflicting Network Slice Requirements	eMBB Slice: Demands large, contiguous spectrum for high throughput. - URLLC Slice: Requires guaranteed, low-latency spectrum with high reliability. - mMTC Slice: Needs sporadic, low-bandwidth connections for massive device counts. - Core Conflict: Real-time, dynamic resource allocation is needed to prevent SLA breaches (e.g., for URLLC) and ensure overall spectral efficiency. [8][32]
Network Densification	Density Challenge: Heterogeneous Networks (HetNets) [26]	Proliferation of low-power small cells (micro, pico, femto) creates a dense network topology. - Leads to a complex and chaotic interference landscape. - It manifests as damaging co-channel and adjacent-channel interference, which might imperil the capacity gains of densification.[7]

actual user data can be sent. The allocation of such important resources may lead to reduced peak throughput and total capacity (especially when compared to a single-carrier solution) [3]. Dynamic scheduling in LTE/NR involves many complex variables with decisions having to be made every 1ms per user depending upon real-time traffic information, the device type, and the fairness of the routing of the traffic. Thus, if there are inaccuracies in any one of the aforementioned parameters, then the performance of the system would be compromised. As such, this represents the greatest technological accomplishment in the development of radio systems as it has become a trade-off.

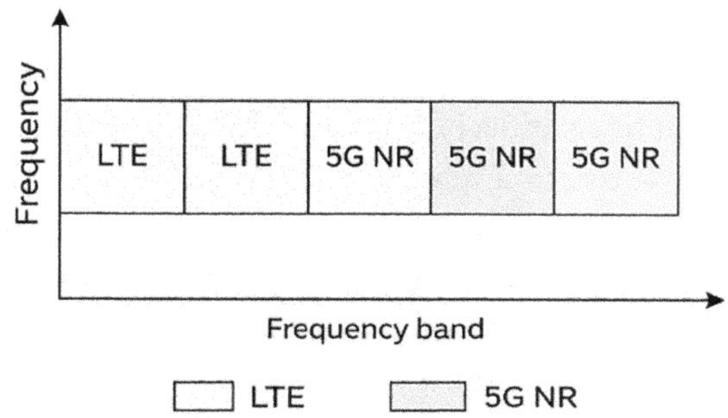

FIGURE 1.7 Dynamic spectrum sharing (DSS). [2], [7]

1.2.6 Blockchain for Decentralized Spectrum Management

The growth of the deployment of wireless devices and introduction of new communication services has heightened the pressure on radio spectrum resources. As 5G networks are rolled out and 6G systems are expected to be developed, the list of devices and applications that require connection and transfer large amounts of data is rapidly expanding. Traditional spectrum management models have been based on centralized regulatory bodies that allocate frequency bands to service providers using long term licensing models. Despite the benefits of this centralized structure in providing regulatory control, it can frequently lead to inefficient spectrum use because of the absence of policies that allow flexibility to dynamic network conditions and therefore, the policies remain unchanged. In life in general, allocated spectrum resources are underutilized in some geographic areas or at some point in time, when other operators are undergoing spectrum crises in some areas. These inefficiencies demonstrate the necessity of more flexible and open spectrum management structures that would allow dynamic allocation and sharing systems. The most recent developments in distributed ledger technologies and specifically blockchain will offer a potential solution to these issues through decentralized communication between various stakeholders that can contribute to the spectrum access and spectrum management. A blockchain technology is a distributed database that maintains an ever-increasing list of records, also known as blocks, connected with the help of cryptographic hashes. Contrary to the classical centralized databases, blockchain networks are decentralized with every transaction being stored on a network of nodes with each node having a synchronized copy of the ledger. The validation of transactions occurs via consensus protocols and only verified information is added to the chain.

The architecture presents inalienable benefits that include data immutability, transparency, and resistance to unauthorized alterations. The latter leads to blockchain being especially applicable to the applications that need secure and auditable resource allocation procedures [28, 29].

In a wireless communications environment, blockchain may be used to enable decentralized spectrum management through securely storing the ownership of the spectrum, access privileges, and history of spectrum usage. Spectrum allocation policies and policies and regulatory policies can be directly encoded into the system through the utilization of smart contracts, programmable scripts that are automatically run on the blockchain. Under predetermined criteria, these smart contracts automatically cause spectrum allocation or revocation events. The mechanism minimizes the use of manual intervention and allows the implementation of spectrum policies automatically.

Decentralized auctions for spectrum make use of blockchain technology to management of the spectrum. It provides transparency and tamper-proof records for all transactions. This reduces conflicts and increases accountability. The role of blockchain becomes even more important in the context of 5G and even 6G networks. This is because many applications are expected to be supported in these networks, such as smart cities, IoT, autonomous transport systems, and industry automation. In these applications, flexible allocation of bandwidth, latency, and reliability may be required. Blockchain marketplaces help in the dynamic trading of the spectrum in accordance with real-time requirements. It also enables network slicing through the secure management of service-level agreements (SLAs) through smart contracts for URLLC, mMTC, and eMBB services. However, there are challenges to be addressed. These challenges are related to the scalability of the blockchain due to the high volumes of transactions, the delay caused by computationally expensive consensus methods, regulatory considerations in line with existing laws regarding the spectrum, and the need to maintain secrecy in the operations while the system inherently provides transparency.

To address these shortcomings, scientists are looking into the concept of hybrid structures which would make use of blockchain-based transparency and regulatory control by a central authority. Lightweight consensus mechanisms, permissioned blockchain networks and integration of edge computing are also under investigation to improve performance and scalability. To conclude, blockchain technology suggests an attractive model of

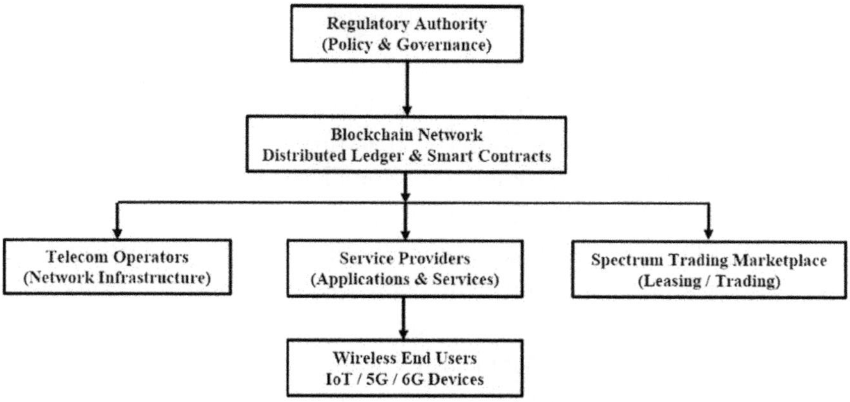

FIGURE 1.8 Blockchain-enabled decentralized spectrum management architecture. [23], [28]

facilitating the decentralized, transparent, and safe spectrum handling in the next generation wireless networks.

The suggested architecture shows a network of a decentralized spectrum management with the help of blockchain. The regulatory authority establishes policies and mechanisms of spectrum access. A blockchain network is a distributed registry storing transaction of spectrum allocation and implementing rules by smart contracts. The blockchain interacts with telecom operators, service providers, and spectrum trading platforms in order to facilitate secure spectrum sharing and dynamic allocation. These network entities ultimately reach wireless users and IoT devices making sure that available radio spectrum is well utilized.

1.3 THE 6G HORIZON: EXPONENTIATION THE CHALLENGES

Although 5G has added complexity to managing the radio frequency (RF) spectrum, the envisioned applications of 6G are anticipated to increase by an order of magnitude or more over what is experienced with 5G. 6G targets such as terabits (Tbps) speeds, sub-millisecond latency, and omnipresent, smart connectivity cannot be attained by merely extending 5G technologies. 6G requires a new, fundamental shift in range of the spectrum, the physics of propagation, and the intelligent core that will govern the network. These dramatic advances will shift the complexity of the problem set into new orders of magnitude, and require cognitive, predictive, and integrated problem solving [30, 31, 32].

1.3.1 The Terahertz (THz) Frontier (0.1–10 THz)

The ever-increasing need for bandwidth has always prompted the advancement of communications to higher frequencies. While 5Gtechnology made use of the millimeter-wave (mmWave) frequencies, 6G technology will utilize the Terahertz (THz) frequencies. The THz frequencies are the mostly unused portions of the electromagnetic spectrum and are between 0.1 and 10 THz. The THz spectrum has an almost unlimited bandwidth, providing large contiguous blocks of spectrum compared to the mmWave spectrum. This will help achieve the 6G technology target of Tbps data transmission. This increased demand on data transfer will include holographic communication, extended reality (XR), and many other applications requiring the creation of nearly instantaneous digital twins in a real time environment where a significant amount of massive datasets need to be transferred and processed almost as quickly [12,13,14].

THz band operation faces problems much larger than the mmWave RF spectrum. Challenges in THz are described below:

I. **Extreme Physics:** THz waves have one of the most complicated sets of physics of all frequencies. This is because at certain distances different transmission windows due to molecular absorption develop. If an atmosphere has water vapor and certain other molecular compounds, the signal becomes blocked and will result in weak transmission. The signal is also blocked by solid objects, so communication can only be done over a distance of a few meters and, more often than not, will require a clear line of sight over several kilometers [14, 15]. The THz systems have to handle very-large-scale antenna arrays in order to produce the very narrow

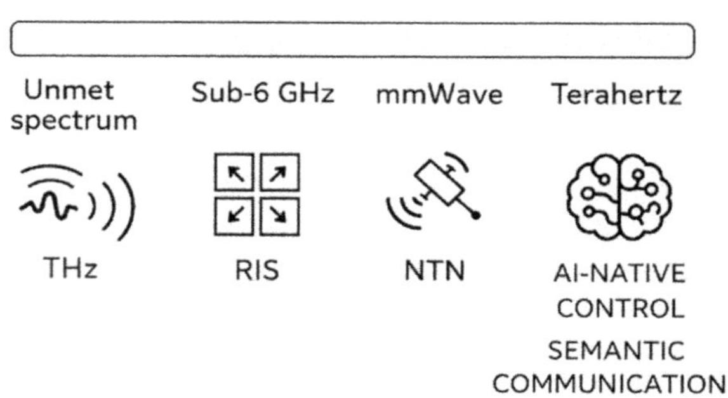

FIGURE 1.9 6G spectrum frontier. [31], [33], [34]

"pencil-beams" that these systems require. The beams are far less than their mmWave equivalents. This makes the initial signal acquisition and beam tracking for mobile users very complicated. The so-called "beam split effect" makes the problem worse, since it causes frequencies in a wideband signal to steer in different directions. This results in complicated alignment of the beam.

II. **Hardware:** The development of THz communication systems and THz technology is hindered by the 'THz gap'. This is the 'gap' between the electronics and photonics industries which requires the development of high functioning systems. Modern semiconductor technologies fail to generate, amplify, and receive THz signals with proper efficiency. The power output of transceivers, which are still in their developmental stage, tends to be considerably low, and there is a high rate of self-generated noise, thus diminishing the transmission's range and reliability. The high power dissipated as a result of this inefficient hardware also creates a significant obstacle to the integration of these systems in mobile, battery-operated devices [14].

III. **Management:** Access to spectrum is fundamentally different due to the presence of such physical and hardware constraints. The frequency and time dimension map is gone, the radio spectrum becomes a high dynamic three-dimensional resource depending on space (direction), distance and LoS availability. Management of such systems needs to support the discovery and scheduling of short-lived, highly directional, distance-dependent spectral windows in real time which is an order-of-magnitude beyond what can be achieved using state-of-the-art spatiallyaware algorithms.

1.3.2 Cell-Free Massive MIMO and Reconfigurable Intelligent Surfaces (RIS)

6G aims to innovate the cellular network model to mitigate the high-frequency range and interference in densely packed networks. To achieve this, two technologies will be integrated: reconfigurable intelligent surfaces (RIS) and cell-free massive MIMO [1, 3, 4,15].

Cell-free massive MIMO technology will improve the cellular network model by eliminating cells and instead having "many distributed access points (APs) be coordinative and uniformly serving all the users in a geographical region," a dramatic enhancement in service quality and cell edge performance. While the multipurpose "Active Metasurfaces" proposed in the RIS technology seem "more radical," they transform a network's "passive and uncontrollable" underlying wireless space into an "active and optimizable" integral part of the configured network. An RIS consists of low-cost passive metasurface components that alter the phase of wireless electromagnetic waves. By shifting phases and reflecting energy in an RIS, one can beam direct wireless energy and interfere at a distance to create "virtual line of sight" paths. All the walls, windows, and billboards of the environment can be utilized as an antenna system [16, 15, 18].

1.3.3 The 3D Network: Satellites, Drones, and HAPS

The goal for 6G is having interconnectedness anywhere and everywhere, including non-ground structures. The intention is to build a comprehensive Non-Terrestrial Network (NTN) that integrates with a terrestrial network (TN) and layers of aerial and extra-terrestrial structures. This "3D network" would be a fully functioning integrated system comprising constellations of Low Earth Orbit (LEO) satellites, High-Altitude Platform Stations (HAPS) situated in the stratosphere, and drone swarms, all collaborating to deliver uninterrupted and worldwide service [28,29].

3D Interference Challenge: The synthesis of these diverse layers results in exceedingly complex 3D superposition problems. When considering the complexities of altitude and movement, static and ground-based interference models become invalid. Take, for example, a single LEO satellite beam. The diameter of its coverage area may exceed 1000km providing potential co-channel interference with hundreds or thousands of terrestrial cells. The inclusion of HAPS and drone networks adds further layers of complication, because the movement of these aerial networks creates volatile and rapidly shifting interference patterns. The ideal method for the control of interference in this 3D multi-layers scenario is the coordination of resource distribution across ground stations, satellites in various orbits, and aerial vehicles which are in motion. These challenges in interference management are of a magnitude and dynamic performance far beyond the reach of current systems in use [21,28,29].

1.3.4 Ubiquitous AI and Semantic Communication

The final step in 6G advancement is the shift to an AI-native network, where AI is more than a network optimization tool. This allows the network to run itself. AI also opens the door to a new communication type: semantic communication [33, 34].

Unlike classical communication explained by Shannon, which aims at perfect duplication of a stream of bits regardless of the message, semantic communication resolves the issue of meaning. Rather than compressing and transmitting a high-res video of a scene, an AI transmitter can zoom in on the scene and extract the critical semantic detail: "a red car is approaching an intersection" and transmit that meaning in a shortened form. The AI-powered receiver will use external contextual information to fill in and reconstruct the scene and understanding. This technique will enable a dramatic decrease in the volume of data that must be sent, offering unprecedented levels of spectrum and energy efficiency [30,31].

Predictive Allocation and New Threats. In the context of an AI-native stage, spectrum management becomes intent-based and predictive. For the first time, the network has to learn and predict user intent to provision spectrum and other resources pre-emptively and before being requested. Moving from reactive to proactive orchestration will be critical to satisfying the ultralow latency targets expected from 6G. Nonetheless, the advanced use of AI poses disproportionate risks. The AI algorithms directing network traffic will become powerful attack surfaces for any adversary. Data poisoning, adversarial attacks, and other means will be available to an adversary wanting to exploit the AI systems to bend, break, or alter decision-making processes for malicious purposes. New AI "spectrum aware" attacks will be possible in the future. For example, an adversary may game the AI systems to gain disproportionate network access or spectrum and deny or interfere with service to legitimate users. Predictive user networks will also raise unprecedented privacy issues, requiring new legislation and privacy technologies [31, 32].

1.4 AUTONOMOUS SPECTRUM ORCHESTRATION USING MACHINE LEARNING

The emergence of technologies of AI and ML will be used to a large extent in the future to run 6G AI-native wireless networks based on real-time spectrum management and autonomous optimization of the network. The growing concentration of wireless devices, non-homogenous service demands, and changing spectrum states render manual or rule-based spectrum allocation unproductive. Community: The AI and ML models are able to analyze massive amounts of network data and make smart decisions relating to dynamic spectrum distribution, interference control, and traffic forecasts [35]. Deep reinforcement learning (DRL) is one of the strongest AI solutions for spectrum management in a completely autonomous manner. As a part of this, intelligent network agents learn to utilize spectrum resources in an optimal manner by interacting with the wireless environment. They analyze various parameters such as spectrum, traffic, interference, and user density in real time to allocate spectrum resources for various services such as enhanced mobile broadband (eMBB), ultra-reliable low latency communications (URLLC), and massive machine type communications (mMTC). It utilizes algorithms such as deep Q-network (DQN) and proximal policy optimization (PPO) to make decisions in a more efficient and flexible manner compared to traditional spectrum management solutions. This ensures better spectrum utilization and lower latency in wireless communication systems. Federated learning (FL) is a learning technique to sense and predict spectrum in a distributed manner. As a part of this, IoT devices and base stations cooperate to learn from data, and they share updates with a centralized server in a distributed manner, ensuring data privacy and lower communication costs [37]. Transfer learning can also be applied in spectrum management applications where there can be a paucity of labeled data. An example is that models that have been trained on urban network environments can be re-trained on rural deployments with limited retraining. Additionally, explainable artificial intelligence (XAI) is gaining relevance in regulatory compliance and disclosure in spectrum allocation. SHAP (SHapley Additive exPlanations) and LIME (Local Interpretable Model-Agnostic Explanations) are the methods that can be used to interpret the predictions made by ML models and assure the network operators and regulators that the automated decisions made by the spectrum are understandable [38]. In general, autonomous, adaptive, and efficient spectrum management, which can be facilitated through the implication of AI and ML into the wireless networks, is one of the primary demands of the future 6G communication systems.

The architecture is a representation of merging AI and ML methods in autonomous spectrum orchestration in future wireless networks. Data on the network based on wireless devices, base stations, and spectrum sensors are handled in a data collection layer which synthesizes data regarding

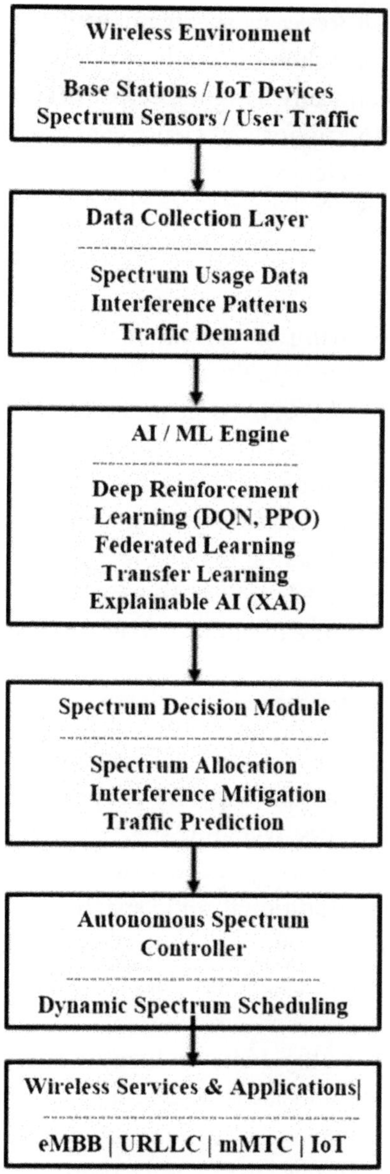

FIGURE 1.10 AI/ML-driven autonomous spectrum orchestration design of 6G wireless systems. [27], [32], [36]

spectrum utilization, the level of interference, and the amount of traffic requests. This information is then fed through an AI/ML engine that uses methods of deep reinforcement learning, federated learning, transfer learning, and explainable AI. According to the acquired models, the spectrum decision unit forecasts spectrum availability and calculates the best resource allocation schemes. Lastly, an independent spectrum controller is used to dynamically assign spectrum capacity to other wireless services to allow it to support efficiently enhanced mobile broadband and ultra-reliable low-latency communications as well as massive IoT applications in 6G networks.

CONCLUSION

The progression of spectrum management reflects a clear transition from rigid allocation systems to intelligent, self-regulating networks. As communication moves into the THz range and integrates with reconfigurable surfaces and non-terrestrial systems, the scale and complexity of spectrum interactions will exceed human control. The next phase is a self-sustaining, predictive network that allocates, adapts, and optimizes spectrum resources in real time. This transformation marks the rise of spectrum intelligence a system that anticipates needs, minimizes interference, and continuously evolves to support the future of global wireless connectivity.

PRACTITIONERS TAKEAWAYS

The necessity for more adaptable, clever, and effective use of radio airwaves is highlighted by the development of spectrum management for 5G and upcoming 6G networks. Dynamic spectrum management techniques, which enable real-time sharing and distribution of spectrum resources, are gradually replacing traditional static spectrum allocation. To increase usage and lower interference, practitioners should use cutting-edge methods including dynamic spectrum sharing, cognitive radio, and spectrum sensing.

In order to forecast traffic demand, identify spectrum availability, and automatically optimize allocation decisions, AI and ML are becoming indispensable tools. Immersion communication, autonomous systems, and enormous Internet of Things (IoT) connectivity are just a few of the new applications made possible by the deployment of higher frequency bands, including as millimeter wave and possibly terahertz frequencies.

Coordinated spectrum management solutions are necessary for the integration of diverse networks, such as small cells, satellite systems, UAVs, and IoT devices. By allowing targeted and service-specific resource allocation, technologies like massive MIMO, beamforming, and network slicing significantly improve spectrum efficiency. Practitioners also need to take into account regulatory structures that facilitate flexible licensing and spectrum sharing.

To guarantee effective, scalable, and sustainable wireless communication networks for 5G and beyond, future-ready spectrum management should include intelligent automation, cutting-edge wireless technologies, and supportive regulations.

REFERENCES

[1] S. Sirasanagandla, "Evolution of 5G Technology for Modern Wireless Communication," *Indian Journal of Scientific Research(IJSR)*, vol. 17, no. 2, pp. 467–473, 2018.

[2] F. Beltrán and M. Massaro, "Spectrum Management for 5G: Assignment Methods for Spectrum Sharing," in *Proceedings of the 29th European Regional Conference of the International Telecommunications Society (ITS)*, Trento, Italy, Aug. 2018.

[3] D. Tan, L. Liu, Y. Zhou, J. Shi, Z. Pan, and H. Wang, "Decentralized Auction for Blockchain-Based Dynamic Spectrum Sharing in 6G Networks," *International Conference on Future Communications and Networks (FCN)*, Valletta, Malta, pp. 1–6, 2024.

[4] M. Vincenzi, A. Antonopoulos, and C. Verikoukis, "Multi-Tenant Slicing for Spectrum Management on the Road to 5G," *IEEE Wireless Communications*, vol. 24, no. 5, pp. 118–125, 2017.

[5] Y. Niu, Y. Li, D. Jin, L. Su, A., and V. Vasilakos, "A Survey of Millimeter Wave Communications (mmWave) for 5G: Opportunities and Challenges," *Wireless Networks*, vol. 21, no. 8, pp. 2657–2676, 2015.

[6] A. Alhammadi, I. Shayea, et al., "Artificial Intelligence in 6G Wireless Networks: A Comprehensive Survey," *IEEE Access*, vol. 11, pp. 54321–54355, 2023.

[7] O. Liberg, C. Hoymann, C. Tidestav, et al., "Introducing 5G Advanced," *IEEE Communications Standards Magazine*, vol. 8, no. 1, pp. 52–57, 2024.

[8] X. Lin, "An Overview of AI in 3GPP's RAN Release 18: Enhancing Next-Generation Connectivity?," *IEEE Communications Society Technology News*, 2024.

[9] J. Xie, "Research on Dynamic Resource Allocation Optimization in 6G Networks Based on Deep Reinforcement Learning," *International Conference on Signal Processing, Communication, and Control Systems (SPCCS 2025)*, vol. 13705, pp. 608–614, 2025.

[10] C. D. Bikkasani, "AI-Driven 5G Network Optimization: A Comprehensive Review of Resource Allocation Traffic Management, and Dynamic Network Slicing," *American Journal of Artificial Intelligence*, vol. 8, no. 2, pp. 55–62, 2024.

[11] S. Chaudhari and H. Kwon, "Machine Learning Based Interference Whitening in 5G NR MIMO Receiver," in *IEEE 95th Vehicular Technology Conference: (VTC2022-Spring)*, IEEE, pp. 1–6, 2022.

[12] N. Li, Q. Sun, L. Wang, X. Xu, et al., "Towards AI-Native RAN: An Operator's Perspective of 6G Day 1 Standardization," arXiv:2507.08403 [cs.NI], Jul. 2025.

[13] T. S. Rappaport, Y. Xing, O. Kanhere, S. Ju, A. Madanayake, et al., "Wireless Communications and Applications Above 100 GHz: Opportunities and Challenges for 6G and Beyond," *IEEE Access*, vol. 7, pp. 78729–78757, 2019.

[14] S. Yao and J. Teng, "Terahertz Communication for 6G Networks: Opportunities and Challenges," *Applied and Computational Engineering*, vol. 46, pp. 232–241, 2024.

[15] S. Thomas, J. Singh Virdi, A. Babakhani, et al., "A Survey on Advancements in THz Technology for 6G: Systems, Circuits, Antennas, and Experiments," *IEEE Open Journal of the Communications Society*, vol. 6, pp. 1998–2016, March 2025.

[16] M. Di Renzo, et al., "Smart Radio Environments Empowered by Reconfigurable AI Meta-Surfaces: An Idea Whose Time Has Come," *EURASIP Journal on Wireless Communications and Networking*, vol. 2019, no. 1, Article 129, May 2019.

[17] E. Shi, J. Zhang, H. Du, B. Ai, C. Yuen, et al., "RIS-Aided Cell-Free Massive MIMO Systems for 6G: Fundamentals, System Design, and Applications," *Proceedings of the IEEE*. vol. 112, pp. 331–364, 2024.

[18] Y. Zhao and X. Lv, "Reconfigurable Intelligent Surfaces for 6G – Applications, Challenges, and Solutions," *Frontiers in Information Technology & Electronic Engineering*, vol. 23, no. 12, pp. 1–18, Dec. 2022.

[19] Z. We, H. Qu, et al., "Integrated Sensing and Communication Signals Toward 5G-A and 6G: A Survey," *IEEE Internet of Things Journal*, vol. 10, no. 13, pp. 11068–11092, 2023.

[20] R. Bennett, "Technical Principles of Spectrum Allocation," in Proceedings of the *41st Research Conference on Communication, Information and Internet Policy, (TPRC 41)*, TPRC Arlington, VA, USA 2013.

[21] J. Mitola and G. Q.Maguire, Jr., "Cognitive Radio: Making Software Radios More Personal," *IEEE Personal Communications*, vol. 6, no. 4, pp. 13–18, Aug. 1999.

[22] G. Ghosh, P. Das, and S. Chatterjee, "Cognitive Radio and Dynamic Spectrum Access – A Study," *International Journal of Next-Generation Networks*, vol. 6, pp. 43–60, 2014.

[23] D. Cuellar, M. Sallal, and C. Williams. "BSM-6G: Blockchain-Based Dynamic Spectrum Management for 6G Networks: Addressing

Interoperability and Scalability," *IEEE Access*, vol. 12, pp. 59643–59664, 2024.

[24] P. Agarwal, M. H. Manekiya, T. Ahmad, et al., "A Survey on Citizens Broadband Radio Service (CBRS)," *Electronics*, vol. 11, no. 23, p. 3985, 2022.

[25] H.-J. Kwon, J. Jeon, and A. Bhorkar, "Licensed-Assisted Access to Unlicensed Spectrum in LTE Release 13," *IEEE Communications Magazine on LTE Evolution*, vol. 55, no. 2, pp. 201–207, 2016.

[26] M. A. S. Choudhury, et al., "Interference Management in Heterogeneous Networks," in *5G- Enabled Internet of Things*, IET, 2021.

[27] K. B. Letaief, Y. Shi, et al., "Edge Artificial Intelligence for 6G: Vision, Enabling Technologies, and Applications," *IEEE Communications Magazine*, vol. 40, no. 1, pp. 5–36, 2022.

[28] K. Kotobi and S. G. Bilen,"Blockchain-Enabled Spectrum Access in Cognitive Radio Networks," *IEEE Network*, vol. 32, no. 3, pp. 28–34, May–Jun. 2018.

[29] Z. Zheng, S. Xie, H. Dai, X. Chen, and H. Wang, "Blockchain Challenges and Opportunities: A Survey," *International Journal of Web and Grid Services*, vol. 14, no. 4, pp. 352–375, 2018.

[30] M. A. Jamshed, et al., "Non-Terrestrial Networks for 6G: Integrated, Intelligent, and Ubiquitous Connectivity," *IEEE Communications Standards Magazine*, vol. 9, no. 3, pp. 86–93, Sept. 2025.

[31] M. H. Alsharif, A. H. Kelechi, M. A. Albreem, et al., "Sixth Generation (6G) Wireless Networks: Vision, Research Activities, Challenges and Potential Solutions," *Symmetry*, vol. 12, no. 4, pp. 676, Apr. 2020.

[32] Q. Cui, et al., "Overview of AI and Communication for 6G Network: Fundamentals, Challenges, and Future Research Opportunities," *Science China Information Sciences*, vol. 68, no. 7, pp. 171301:1–171301:61, 2025.

[33] G. Geraci, D. López-Pérez, M. Benzaghta, et al., "Integrating Terrestrial and Non-Terrestrial Networks: 3D Opportunities and Challenges," *IEEE Communications Magazine*, vol. 61, no. 4, pp. 42–48, April 2023.

[34] Y. Wang, H. Han, Y. Feng, J. Zheng, and B. Zhang, "Semantic Communication Empowered 6G Networks: Techniques, Applications, and Challenges," *IEEE Access*, vol. 13, pp. 28293–28314, 2025.

[35] M. Chen, U.Challita, W. Saad, C. Yin, and M. Debbah, "Artificial Neural Networks-Based Machine Learning for Wireless Networks: A Tutorial," *IEEE Communications Surveys & Tutorials*, vol. 21, no. 4, pp. 3039–3071, 2019.

[36] ITU-R, *IMT Vision – Framework and Overall Objectives of the Future Development of IMT for 2020 and Beyond*, Recommendation ITU-R

M.2083-0, International Telecommunication Union, Geneva, Switzerland, Sep. 2015.

[37] Q. Yang, Y. Liu, T. Chen, and Y. Tong, "Federated Machine Learning: Concept and Applications," *ACM Transactions on Intelligent Systems and Technology*, vol. 10, no. 2, pp. 1–19, 2019.

[38] S. Lundberg and S. Lee, "A Unified Approach to Interpreting Model Predictions," in *Proceedings of the Neural Information Processing Systems (NeurIPS)*, Curran Associates, Inc., 2017.

CHAPTER 2

5G and 6G Communication Systems

Architectural Frameworks and Next-Generation Enabling Technologies

Prachi Pol, Akhilesh Kumar,
Radhika D. Joshi, and Shubhangi Deshmukh

2.1 INTRODUCTION

Over the past many decades, a lot has changed about how to make phone conversations without wires. 5G and the next 6G model are replacing first-generation (1G) analogue cell phones with digital networks that are faster and smarter. At originally, systems could simply make calls and didn't have a lot of space for data. But when 3G and 4G systems got digital modulation, packet switching, and broadband wireless technology, mobile communication became a rapid, data-driven world. The rules for 5G networks were created by the 3rd Generation Partnership Project (3GPP). These networks have made wireless significantly better by making it possible to have improved Mobile Broadband (eMBB), Ultra-Reliable Low-Latency Communication (URLLC), and massive Machine-Type Communication (mMTC) [1]. These kinds of services could revolutionise a lot of things, such self-driving cars, smart cities, connected healthcare, smart transportation systems, immersive extended reality (XR), and industrial automation. Some of the new technologies that come with 5G are beamforming, network slicing, massive MIMO, and the ability to work in millimetre-wave (mmWave) bands. This usually makes the network perform better and faster.

DOI: 10.1201/9781003731597-2

There are now a lot more Internet of Things (IoT) devices, AI-powered apps, edge computing platforms, robotic aerial systems, and high-definition multimedia apps. Because of this, the requirement for wireless bandwidth has skyrocketed. Billions of gadgets will be able to connect to the next generation of 6G networks, which will cause traffic levels that have never been seen before. These ecosystems will need to be very reliable, have a lot of connections, and use less energy. They will also need to have very low latency. The new 6G systems will work in both the terahertz (THz) and millimetre wave (mmWave) frequency bands, as planned. They can send data at speeds of up to 1 Tbps, and it only takes a few microseconds to get there [2]. Higher frequency bands have a lot greater bandwidth, but they also have more technical problems, like big transmission loss, hardware limits, molecular absorption effects, and channels that vary very quickly.

The Federal Communications Commission (FCC) and other regulatory bodies have determined that specific frequency bands are underutilised in certain locations and periods [3]. Standard norms for assigning spectrum use a tight and limited licensing structure. This means that some companies or services always get specified frequency bands. This system of restrictions stops unwanted interference, but approved users don't always use the bands they are given, which is negative for spectrum use. Now, wireless systems are in a strange situation: the frequency looks to be running out, yet a lot of it isn't being used at different times and places. There are a number of innovative and adaptable approaches to use the airwaves to solve this problem. Joseph Mitola [4] was the first person to think of cognitive radio (CR), which is one of the most fundamental ideas in this field. CR is a smart wireless speaker system that can hear what's going on around it, remember what it has learnt, and modify its broadcast settings right away, such as the carrier frequency, bandwidth, modulation method, and transmission power. White spaces or spectrum holes are names that people sometimes use for vacant spectrum bands. CR discovers these bands and uses them to get more out of the overall spectrum. This also protects licensed (primary) users from meddling that could harm them [5]. Finding the right frequency is one of the most critical components of CR. It checks to determine if a main user is using a specified group of frequencies right now. Spectrum sensing has to work well so that there isn't as much interference and both primary and secondary customers get good service. identifying energy, identifying traits that match a filter, and finding patterns that happen over and over again are all parts of sensing. A number of people have utilised them and studied them. But they might not perform as effectively when the signal-to-noise ratio (SNR) is poor or when the wireless environment is always changing. AI and machine learning (ML) are being used more and more in systems that watch over and control the spectrum to overcome these problems [6]. If they regard spectrum occupancy detection as a pattern recognition or classification problem, AI-based systems can learn from past data, anticipate when the spectrum will be free, and decide who can use the spectrum before it happens. AI will fuel communication systems that can run and upgrade themselves in 6G networks. These smart systems will likely form a large part of those networks.

In wireless communication engineering, it is a huge deal that the way spectrum is used has changed from being fixed to being flexible and smart. To satisfy the high-performance standards of future 6G networks, we will need to deploy cognitive radio, better ways to watch the spectrum, distributed intelligence, and data-driven spectrum management strategies all at once. Even if wireless systems are getting smarter, connecting more devices, and getting faster, making the optimum use of the spectrum will still be a key aspect of the designs for next-generation communication [1–5].

The heterogeneous network architecture supporting next-generation wireless communication systems is illustrated in Figure 2.1.

2.2 CORE NETWORK ARCHITECTURE OF 5G TECHNOLOGY

2.2.1 Overview

5G is the fifth generation of wireless communication systems standardised by the 3rd Generation Partnership Project (3GPP). It represents a significant technological leap beyond 4G LTE by introducing a flexible, service-oriented architecture capable of supporting diverse application scenarios with vastly different performance requirements. Unlike previous generations that primarily focused on improving data rates, 5G has been designed as a unified connectivity platform enabling enhanced broadband services, mission-critical communication, and massive-scale mac.

Figure 2.2 illustrates the 5G network architecture showing the interaction between the User Equipment (UE), gNodeB, and core network functions such as AMF, SMF, UDM, AUSF, PCF, NSSF, and UPF for session management, authentication, policy control, and data transfer to the Internet.

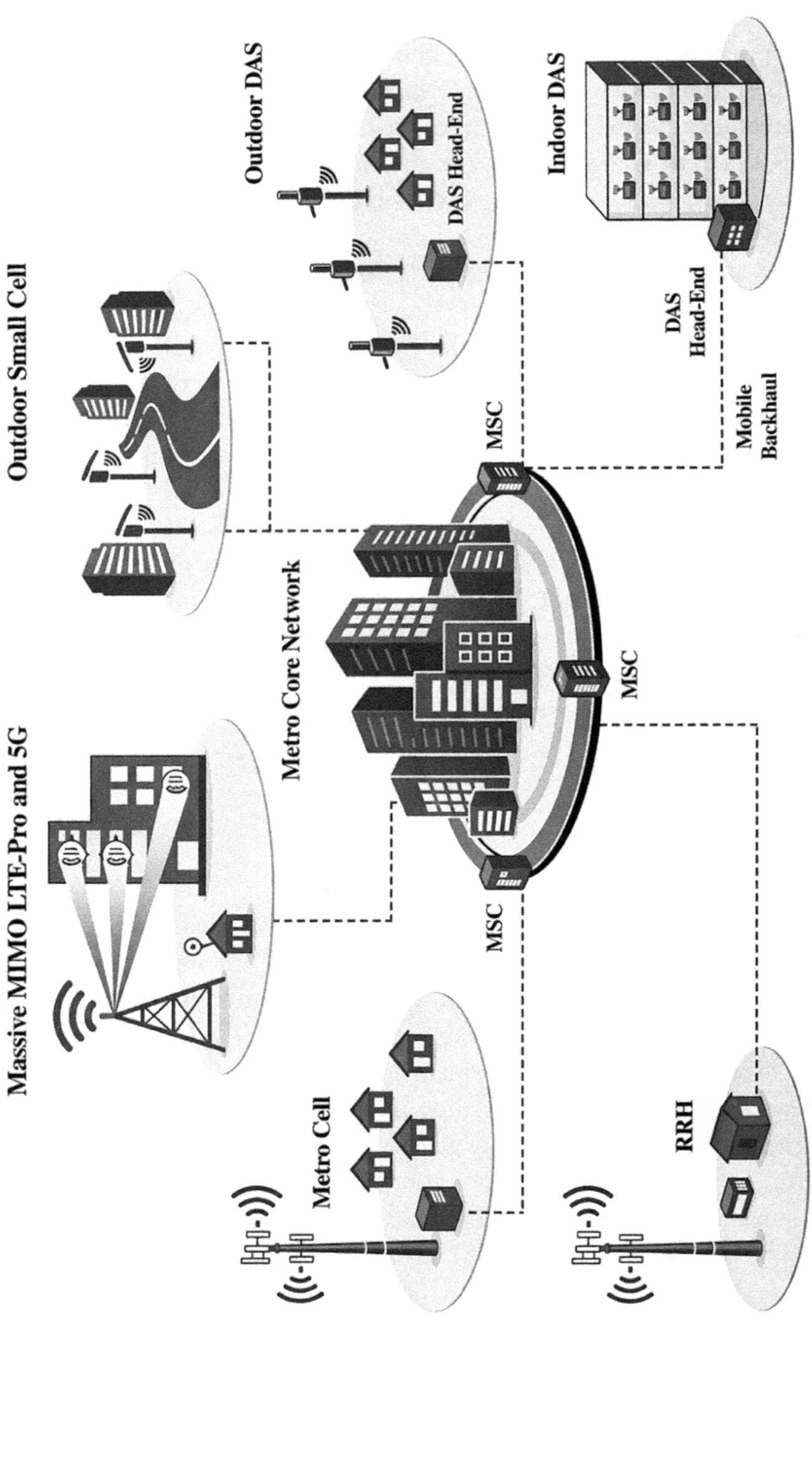

FIGURE 2.1 Heterogeneous network architecture for 5G and 6G systems.

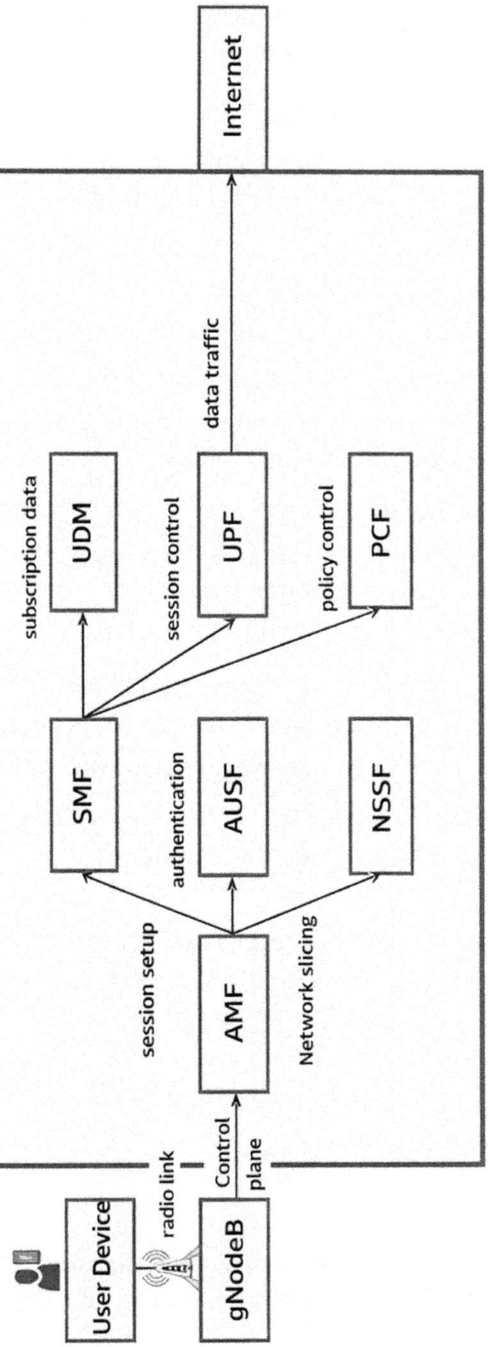

FIGURE 2.2 5G core network architecture.

2.2.2 Key Features of Core Network Architecture in 5G and Beyond

5G introduces several advanced technological features that significantly improve network performance:

- **Data rates up to 10 Gbps:**

  Achieved through wider bandwidth allocation, higher-order modulation schemes, and advanced coding techniques.
- **Latency below 1 ms:**

  Enabled by shorter transmission time intervals (TTIs), edge computing integration, and optimised protocol stack design.
- **Massive MIMO (Multiple Input Multiple Output):**

  Employs a large number of antenna elements at the base station to increase spectral efficiency and spatial multiplexing gains.
- **Beamforming:**

  Focuses transmitted energy toward specific users, improving signal strength and reducing interference.
- **Network Slicing:**

  Allows the creation of multiple virtual networks over a shared physical infrastructure, each optimised for specific service requirements.
- **Millimetres Wave (mm Wave) Communication:**

  Utilises higher-frequency bands to access large bandwidth resources, enabling extremely high throughput.

These technologies collectively enhance system capacity, energy efficiency, and user experience compared to previous generations [7–10].

2.2.3 5G Spectrum Bands

To achieve high data rates and capacity improvements, 5G operates across multiple spectrum ranges:

- **Sub-6 GHz Bands:**

  Provide broader coverage and better penetration characteristics. These bands are suitable for nationwide deployments and urban macro-cell coverage.

2.2.4 Limitations and Challenges in 5G and Beyond Networks

Despite the fact that there have been significant technological advancements achieved in the realm of 5G networks, there are still a great deal of business, technical, and financial concerns that need to be resolved. 5G makes greater use of the spectrum, has a more flexible numbering system, and has the ability to provide virtual networks. But because it is used by such a large number of people, particularly in areas that are really busy, it has revealed a number of issues that are present in the real world and need to be fixed if performance is going to continue to improve over time.

The fact that there are so many people in the area makes it difficult. In particular, this is the case below 6 GHz, which is where the demand for high-speed wireless internet is becoming increasingly urgent. A worsening of frequency occupancy occurs when an increased number of smartphones, IoT devices, and applications that require a significant amount of data are utilised. Because of this, cities are forced to contend with increased levels of competition and a decline in the quality of service (QoS) [10]. As a result, mmWave bands offer a greater variety of bandwidth options than microwave bands do. On the other hand, they have issues with propagation because they lose a significant amount of signal along the way, they are easy to block, and they do not travel as far on average. The only way to cope with interference is for them to be grouped together in very small and tight groups. When it comes to ultra-dense networks (UDNs), co-channel crosstalk is another issue that can arise. There is a greater likelihood of interference occurring in common frequency bands when there is a large number of tiny cells, heterogeneous networks (HetNets), and device-to-device (D2D) transmission mechanisms. The process of coordinating interference, dynamically assigning spectrum, and optimising across layers is now being investigated by researchers [11].

The fact that 5G technologies do not consume a great deal of power is one of the reasons why they are so widely used. Two technologies, massive MIMO and beamforming, need extremely large antenna grids and powerful equipment for signal processing. Both of these technologies are discussed more below. All of these different kinds of technology require a great deal more electricity from base stations. Because there are a great number of gNodeBs and edge computing hubs, the amount of energy that they require is increased. Sustainable "green communication" solutions are essential to achieving success over the long term. Base stations that are powered by renewable energy [12], sleep modes that are functional, and timing that is determined by the amount of energy that is consumed are all available solutions.

A significant amount of effort is required in order to manage services and alter the way resources are utilised. In order to function effectively, various types of services require varying amounts of bandwidth and the ability to slice the network in novel ways. In order to move data quickly, eMBB requires a large amount of data, URLLC requires an incredibly short amount of time to reply, and mMTC requires a large number of connections. For the purpose of ensuring that these services collaborate in real time in the most effective manner possible, we require programmable network control frameworks and optimisation tools that are powered by artificial intelligence [13].

The following are the key areas of technology that are widely discussed as potential solutions to these problems-

- The utilisation of the spectrum in a strategic manner that makes use of both cognitive and dynamic access methods
- Optimisation of the network using artificial intelligence for the purpose of pooling resources and determining how traffic will flow
- Instruments for the management of networks that are capable of being altered and programmed

It is seen that these fundamental ideas will evolve and become fundamental components of wireless systems that are of the sixth generation (6G) [10, 13].

2.3 6G NETWORK ARCHITECTURE WITH AI-ENABLED MASSIVE MIMO COMMUNICATIONS

2.3.1 Overview of 6G Network Architecture with AI-Enabled Framework

6G is the next step in the evolution of wireless communication. Businesses should be able to use it by 2030. 5G generally speeds up connections, but 6G wants to bring together intelligence, sensing, communication, and processing into one cyber-physical ecosystem. It sees an AI-native architecture in which intelligence is built into every layer of the protocol stack, rather than being a separate tool for optimisation [14]. The Figure 2.3 illustrates a 6G network ecosystem integrating ultra-massive MIMO, AI-driven

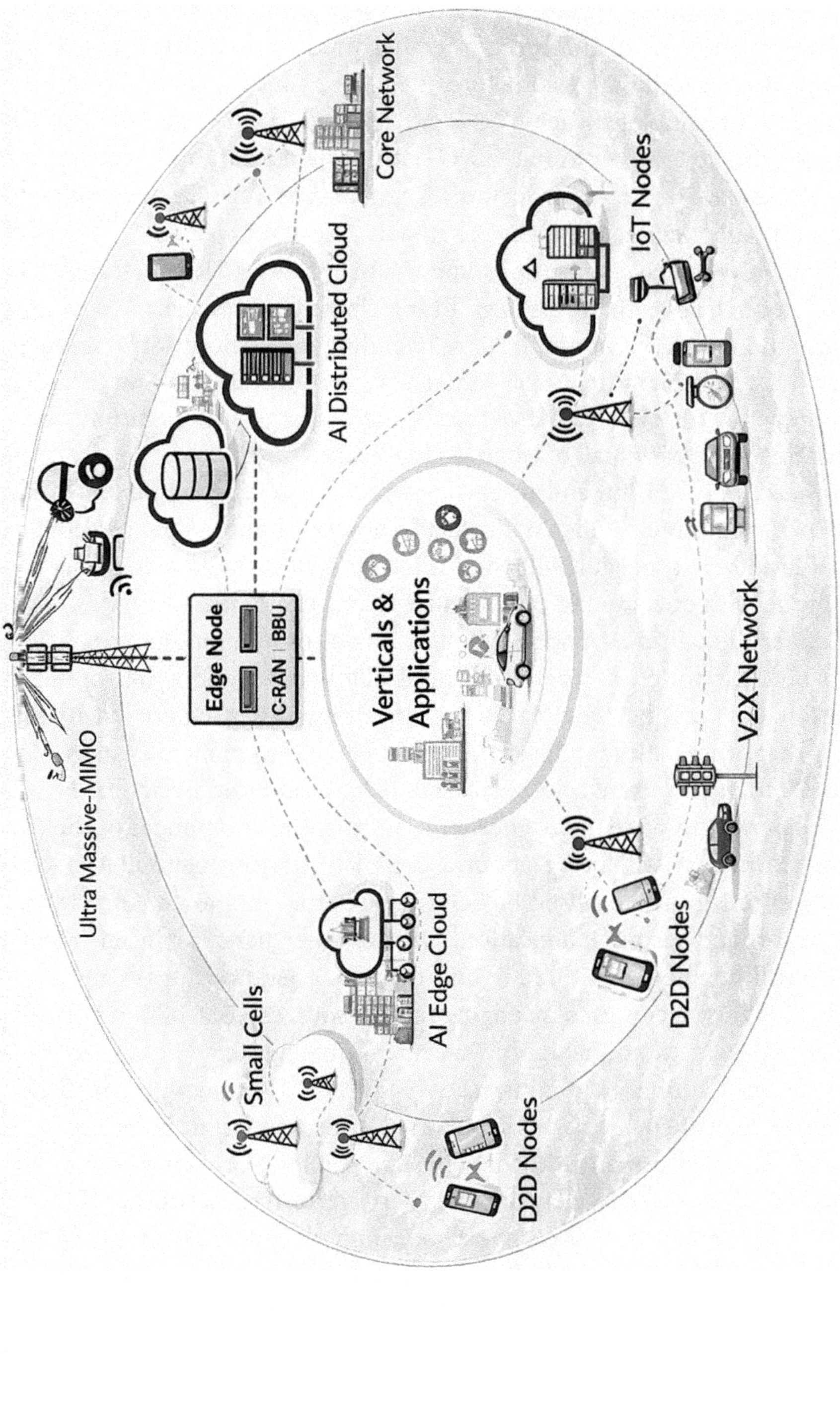

FIGURE 2.3 6G network architecture with AI-enabled edge intelligence and ultra-massive MIMO.

distributed cloud, edge computing, small cells, IoT nodes, D2D/V2X communication, and vertical applications for intelligent and ultra-connected services. 3.2 Features of 6G is planned to work in higher frequency bands, such as the terahertz (THz) and sub-terahertz (sub-THz) ranges. There will be more bandwidth than ever before. It will also connect all kinds of communication technology, such as ground, air, and satellite, into one global network that covers everything. It will be important for the 6G system to be able to keep an eye on the environment in real time, manage the spectrum efficiently, and have edge intelligence that is spread out.

Immersive metaverse settings and virtual reality (XR) environments [14]. Nevertheless, there are significant challenges associated with the propagation of THz communication, including molecule absorption, high route loss, and air attenuation. Therefore, very advanced beamforming techniques, ultra-massive MIMO arrays, and adaptive propagation compensation systems will all be of great significance.

In addition, ultra-low end-to-end latency is an essential component of 6G networks. It is anticipated that it will fall below one hundred microseconds. With this level of speed, it will be possible for systems and devices to communicate with one another in a manner that is nearly instantaneous. As a result, technological advancements such as real-time autonomous control, distributed robotics, accurate industrial automation, and cyber-physical coordination will become feasible applications. When compared to the delay that occurs at the millisecond level in 5G networks, this responsiveness at the microsecond level represents a significant improvement.

The paradigm of AI-native design is an essential component of the 6G architecture that will bring about a complete transformation. Artificial intelligence is mostly utilised by 5G as an external tool to enhance traffic prediction and resource allocation. On the other hand, artificial intelligence will be integrated directly into the core operations of the network with 6G. When it comes to managing radio resources, controlling mobility, enforcing security, and assigning spectrum, intelligence will be included into the procedures [15]. This technology, which is native to artificial intelligence, will enable networks to learn, heal, and build themselves up, which will result in networks that are significantly more reliable and versatile. Mixing terrestrial networks with non-terrestrial networks (NTNs), such as low-earth-orbit (LEO) satellites, high-altitude platforms (HAPs), and unmanned aerial vehicles (UAVs), is another way that 6G is expected to connect people all over the world in three dimensions. This will be accomplished by mixing different types of networks. It is because of this integrated architecture that there will be seamless coverage all throughout the world, particularly in regions that are difficult to access or do not receive sufficient service.

2.3.2 Intelligent Reflecting Surfaces (IRS)

Will also play a significant role in the enhancement of spectral efficiency and coverage. The phase shifts can be altered by these programmable meta-surfaces, which also have the ability to change the way electromagnetic waves move in real time. With the use of IRS technology, transmissions can be made more powerful, interference can be reduced, and the spectrum can be utilised more effectively, all without the need for more active transmission power. Because of all of these aspects, it is clear that 6G is not merely a slight improvement in performance; rather, it is a comprehensive architectural shift toward communication systems that are intelligent and aware.

2.3.3 Technological Drivers and Application Requirements in 6G

One of the primary reasons for the transition to 6G networks is the emergence of new applications that require more power than what can be provided by 5G systems. The next generation of communication frameworks needs to be able to manage speeds, dependability, and intelligence that have never been seen before. This is because digital transformation is accelerating in many different industries. Due to the fact that it transmits three-dimensional holographic content in real time, holographic communication is one of the most challenging applications. It is necessary for these kinds of systems to have data speeds of terabits per second and relatively low latency in order to provide consumers with a seamless and immersive experience. It is impossible for standard broadband technologies to handle this amount of data.

One more significant cause for the creation of 6G is the expansion of the Tactile Internet. Applications like as robotic surgery, remote control of industrial machinery, and immersive virtual reality are all examples of applications that require communication that is extremely reliable, with a latency of less than a microsecond and jitter that is nearly non-existent [14]. These applications could put both safety and performance at risk if they experience delays or lose packets; hence, deterministic communication protocols are an absolute necessity.

Figure 2.4 presents the 6G AICT framework highlighting immersive communication, integrated AI and communication, hyper-reliable low-latency communication, massive and ubiquitous connectivity, and integrated sensing with sustainability, security, and intelligence as core design principles The development of smart cities and Industry 5.0 necessitates the utilisation of sensing and communication capabilities that are compatible with one another. Large-scale automation, collaborative robots, systems for human-machine interaction, digital twins, and other forms of automation all require networks that are able to evaluate sensory data, make decisions based on artificial intelligence, and deliver results in real time.

Additionally, in the future, self-driving automobiles and intelligent transportation systems will require communication networks that are both highly reliable and widely dispersed. It is necessary for vehicle-to-everything (V2X) communication to function with extremely low latency and extremely high dependability in order for individuals to collaborate on decision-making, prevent crashes, and manage traffic in a manner that is beneficial to all parties involved. These various application domains

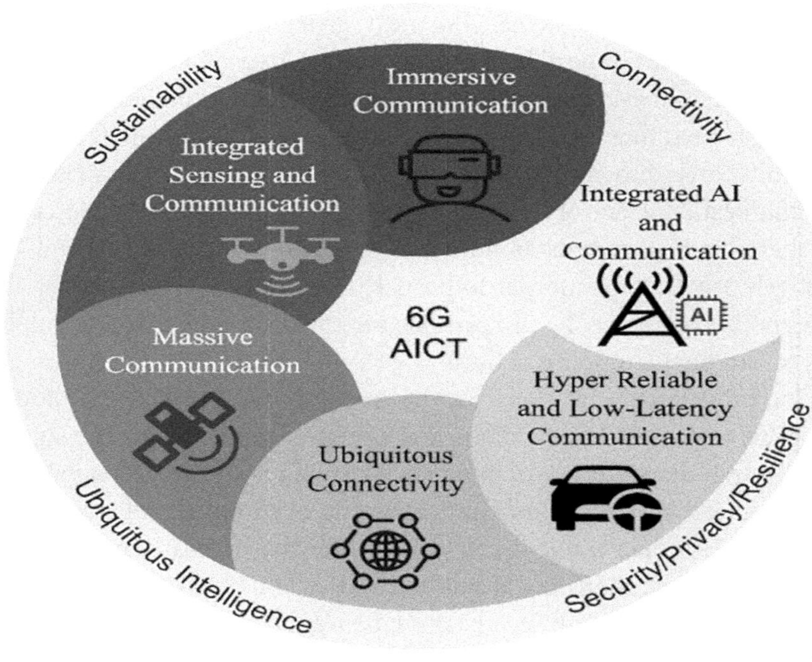

FIGURE 2.4 6G AI-Centric Communication Technology (6G AICT) framework.

demonstrate that we require a communication system that is capable of performing more than merely transmitting data. In addition to this, it must be able to perceive its surroundings, intelligently process information, and adapt to its surroundings on its own. These characteristics are included into the 6G architecture idea by means of AI-native control, integrated sensing and communication (ISAC), and intelligent spectrum management frameworks [15].

An Analysis of the Differences Between 5G and 6G over time

- An examination of the differences between 5G and 6G technologies finds that the former represents a substantial paradigm shift, while the latter is more of a generational improvement.
- With the utilisation of THz and sub-THz frequency bands, it is anticipated that 6G would be able to achieve speeds of up to 1 Tbps, while 5G is capable of reaching rates of up to 10 Gbps.
- Approximately one millisecond is the amount of time that is required for data to travel from one end to the other when using 5G. On the other hand, the goal of 6G is to reduce that time to less than one hundred microseconds, which will make it possible to do autonomous and haptic operations in real time.
- Although 5G operates mostly in the sub-6 GHz and mmWave bands, 6G extends its frequency range all the way up to the terahertz spectrum, which significantly improves the amount of bandwidth that is accessible.
- When it comes to network intelligence, 5G makes use of AI-assisted optimisation, but 6G operates with an architecture that is entirely based on AI, with intelligence being integrated into each and every tier of the network.
- When it comes to coverage, 5G is primarily concerned with land-based networks, but 6G envisions networks that integrate land-based and satellite networks in order to provide a continuous link across the entire planet.

The distinctions between the two make it abundantly clear that 6G is not merely a minor enhancement to the capabilities of 5G; rather, it is a significant change that will result in wireless ecosystems that are intelligent, perceptive, and ultra-high capacity [14], [15].

TABLE 2.1 Comparative Analysis of 5G and 6G Table 2.1 illurates the various of the parameters for 5G and 6G.

Parameter	5G	6G (Expected)
Deployment Period	~2020	~2030
Peak Data Rate	Up to 20 Gbps	Up to 1 Tbps
User Data Rate	~100 Mbps	>1 Gbps
Latency	~1 ms	<0.1 ms
Frequency Bands	Sub-6 GHz, mmWave	mmWave + Terahertz (THz)
AI Integration	AI-assisted	AI-native
Reliability	99.999%	99.99999% (target)
Mobility Support	Up to 500 km/h	>1000 km/h
Network Architecture	Cloud-based	AI-driven, edge-intelligent
Use Cases	Smart cities, IoT, AR/VR	Holographic communication, Digital twins, XR, Tactile Internet
Sensing Capability	Limited	Integrated sensing & communication
Energy Efficiency	Improved over 4G	Ultra-low energy, green networking

2.3.4 Comparative Analysis of 5G vs 6G Networks

In a short length of time, wireless communication has come a long way, going from 1G analogue systems to smart data-based networks. 5G is much better than 4G since it enables you use enhanced mobile broadband (eMBB), ultra-reliable low-latency communication (URLLC), and massive machine-type communication (mMTC). But holographic communication, extended reality (XR), digital twins, autonomous systems, and AI-driven networks need even more power.

2.4 SPECTRUM MANAGEMENT

2.4.1 Definition and Conceptual Framework

Spectrum management, which includes the organised control, distribution, coordination, and monitoring of radio frequency (RF) spectrum resources, is accountable for these resources. Spectrum management is subject to radio frequency (RF) spectrum assets. The objective of spectrum management is to guarantee that these resources operate without interference and are exploited to the fullest extent of their capabilities. In light of the fact that the electromagnetic spectrum is a limited and major natural resource, it is absolutely necessary to have efficient spectrum control in order to make it possible to provide a wide range of wireless services. Among these services are mobile broadband, broadcasting, satellite communication, connections to the IoT, and applications for the military. The vast majority of the time, national regulatory bodies will assign spectrum bands to specific services and operators in order to drastically reduce the amount of interference that is not wanted. Spectrum management, on the other hand, moved from being a primarily regulatory role to becoming a technologically complex subject that incorporates real-time monitoring, dynamic allocation, and intelligent optimisation approaches [16]. Spectrum management has evolved from being a strictly regulatory function. This occurs as a result of the rapid expansion of wireless devices and applications that need for high data rates during their operation. In today's 5G and next 6G ecosystems, spectrum management is a combination of policy frameworks, software-defined networking (SDN), artificial intelligence-driven analytics, and CR approaches [17]. These are all methods that are used to control spectrum. By doing so, it is feasible to make the most effective use of the spectrum while also guaranteeing that the quality of service achieved is of the best possible grade.

2.4.2 Traditional Spectrum Allocation Mechanisms

The notions of long-term licensing and fixed frequency allocations are the primary foundations upon which traditional spectrum allocation systems are built. Within the scope of this system, regulatory agencies provide licensed operators access to particular frequency bands. This enables the operators to operate exclusively within particular geographical areas and for particular amounts of time.

It is vital to note that the following are some of the most significant components of traditional spectrum allocation:

- Fixed allocation: In the context of the field of telecommunications, the term "fixed allocation" refers to the practice of assigning particular frequency bands to particular services, such as cellular, broadcasting, or satellite, regardless of the manner in which these services are currently being employed.
- Licensed Bands: The only way for operators to receive the license to operate in specific spectrum bands is via the use of auctions or the authorisation of the government. This is the only way that operators can obtain this authorisation.
- Static Assignment: In the process of static assignment, spectrum resources are held for the license holder, even when those resources

are not being utilised to a considerable degree. This ensures that the license holder has access to the spectrum resources.

Even though this methodology guarantees constant interference control and regulatory stability, substantial measurement initiatives have demonstrated that permissible spectrum bands are underutilised in a major way both temporally and spatially [16]. This is the case despite the fact that this methodology guarantees both of these things. Certain frequency bands are utilised in metropolitan regions, whereas others are not utilised during off-peak hours or in rural areas. These frequency bands are occupied in urban areas. Due to the fact that this imbalance exists, the spectrum is utilised in an inefficient manner, which calls for the introduction of allocation systems that are more adaptable. The Dynamic Spectrum Management (DSM) system represents a substantial change toward the utilisation of spectrum in real time based on demand. This is in contrast to the static allocation system, which is used to allocate spectrum. Adjusting the frequency in line with the level of interference, the amount of traffic on the network, and the requirements of the service is something that may be accomplished through the utilisation of DSM. When it comes to the DSM, the following are some of the most important concepts:

- Real-Time Allocation: Spectrum resources are assigned on the fly based on the amount of demand that is present on the network as well as the weather. This is done in accordance with the real-time allocation system.
- Opportunistic Spectrum Access: As a result of opportunistic spectrum access, secondary users are allowed to make use of vacant spectrum bands, which are commonly referred to as spectrum gaps, for a limited amount of time without causing any harm to prime users.
- AI-Driven Optimisation: A form of optimisation that is powered by artificial intelligence comprises the utilisation of ML algorithms for the purpose of analysing traffic patterns, determining when there may be congestion, and determining the most efficient ways to reuse frequencies and schedule them.

The term DSM is closely linked to CR, software-defined radio (SDR), and artificial intelligence-native control architectures in networks that make use of 5G and future capabilities. DSM not only makes it feasible to deploy a large number of devices in very congested network conditions, but it also considerably improves the spectral efficiency of the network [17], [18]. The utilisation of intelligent sensing and flexible resource allocation are the means by which this objective is attained.

2.4.3 Spectrum Sharing Models in CR Systems

Several alternative strategies for spectrum sharing have been put into place in order to solve the lack of spectrum and to make better use of it. This was done in order to make the most of the spectrum that is available. When it comes to accomplishing the objectives that they have set out to achieve, these models make an effort to strike a balance between rigid rules and adaptable ways.

- **Licensed Spectrum (Primary Users):**

  A given operator is only awarded the exclusive right to use a specific frequency band if that operator has licensed spectrum. This is the only circumstance under which this privilege is provided. It is likely that this model is not exploited to its full extent during times when there are not a lot of people using it. This could prevent it from reaching its maximum potential. However, it does ensure that interference management is executed in a reliable manner. On account of the fact that it is dependable and guarantees a high level of service, licensed spectrum continues to be the component that is considered to be the most significant in the design of cellular networks [16].

- **Unlicensed Spectrum (Secondary Users):**

  The usage of unlicensed spectrum bands, such as those used for Wi-Fi and Bluetooth, makes it possible for a large number of users to connect, given that they comply with certain laws regarding power and technology. Although this paradigm encourages creative thinking and makes quick deployment easier, it also has the potential to result in congestion and interference that cannot be managed in regions where there are a significant number of people. This is especially true in areas where many people are present.

- **Shared Spectrum:**

  In order to enhance the functionality of objects, shared spectrum models combine parts of licensed and unlicensed frameworks. This is done in order to achieve the objective of improving functionality. As some examples are: Licensed Shared Access (LSA) system Licensed users are able to share spectrum with incumbents through the LSA system, which is in conformity with the regulations that have been established by the government. The Citizens Broadband Radio Service (CBRS) is a service that contains a tiered access structure that enables different user classes to share spectrum in real time while being regulated by centralised mechanisms [18].

2.5 CR

CR, also referred to as CR, is a way of broadcasting and receiving wireless signals that is more intelligent than traditional methods. It makes it possible to make adjustments to the parameters of the transmission on the fly, taking into account the fluctuations in the conditions of the radio environment. In contrast to CR systems, conventional radio systems do not make use of established frequency bands during transmission. This is one of the things that differentiates them from CR systems. As an alternative, they continuously monitor how the spectrum is being utilised, search for frequency bands that are not being utilised (also known as spectrum holes or white spaces), and make adjustments to aspects such as carrier frequency, transmission power, modulation scheme, and bandwidth in order to achieve optimal performance while minimising the risk of harmful interference. This is done in order to achieve optimal performance while minimising the risk of interference.

During the year 1999, Joseph Mitola III was the one who initially conceived of the idea of CR. Through his creative thinking, he conceived of a radio system that was able to perceive its surroundings, acquire knowledge about them, and make adjustments to itself depending on those observations [19]. After some time had elapsed, the official definition of CR stressed the capability of the technology to evaluate the spectral environment and make adjustments to transmission settings on the fly in order to make more efficient use of the spectrum [20]. This was done in order to make the most of the spectrum. CR is now widely regarded to be a crucial component of dynamic spectrum access (DSA), particularly in circumstances when licensed spectrum resources are not being exploited to their full capacity. This is especially true in cases where licensing restrictions are in place. Figure 2.5 illustrates dynamic spectrum access in CR networks where primary and secondary users (SUs) share

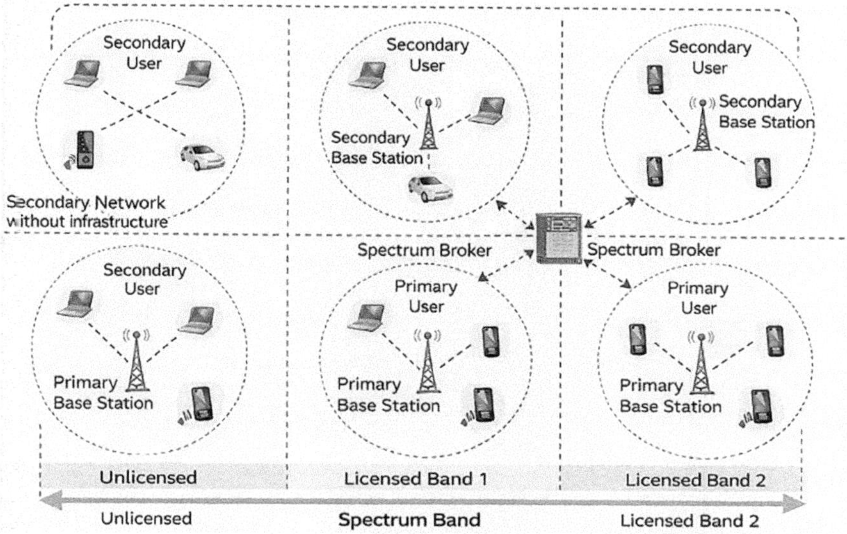

FIGURE 2.5 Cognitive radio spectrum sharing with spectrum broker in licensed and unlicensed bands.

licensed and unlicensed bands under the coordination of a spectrum broker. CR technologies are of uttermost significance in contemporary wireless ecosystems, particularly in the context of 5G and 6G networks that are now in the process of being created. They make it feasible to distribute spectrum in a flexible manner, contribute to the decrease of spectrum scarcity, and give support for resource allocation systems that are driven by artificial intelligence [21].

Figure 2.6 illustrates the CR cycle showing spectrum sensing, spectrum management, spectrum sharing, and spectrum mobility interacting with the radio environment to enable adaptive and intelligent spectrum utilisation. Going through a cognitive cycle that includes sensing, assessing, choosing, and adjusting to external variables is the process that is used to operate a CR system. This cycle is what allows the system to function properly. The following is a list of the major components of CR that are accountable for the functionality described in Figure 2.6.

2.5.1 Spectrum Sensing

First, the identification of the full spectrum the basic role of CR is the ability to see and differentiate between the spectrum. In order to identify frequency bands that are not being utilised to their full potential, it

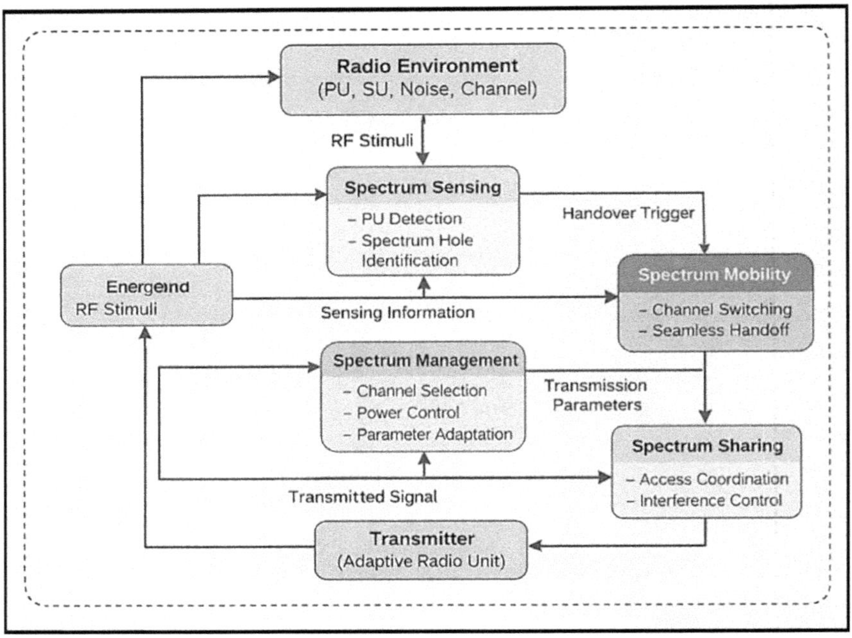

FIGURE 2.6 Functional architecture of cognitive radio network.

is possible to do so by observing signal properties such as energy levels, cyclostationary features, or matching filter responses. The purpose is to identify gaps in the spectrum without causing any inconvenience to those individuals who are permitted to make use of them. It is of the utmost importance to have accurate sensing in order to get rid of interference and provide dependable access to chances that present themselves [20].

2.5.2 Spectrum Decision

Choosing Which Spectrum to Make Use: Following the identification of channels that are available for transmission, the CR takes into account a variety of characteristics in order to determine the frequency band that is the most suitable for transmission. The signal-to-noise ratio (SNR), the level of interference, the bandwidth that is available, and the quality of service (QoS) needs are some of the characteristics that are included in this category. When it comes to spectrum judging systems, it is common practice to make use of optimisation and ML in order to obtain the most possible quantity of data while yet preserving fairness [21].

2.5.3 Spectrum Sharing

Through spectrum sharing, it is possible for several SUs to simultaneously access the same spectral environment on the same network. This allows for several users to share the same spectrum at the same time. Techniques for controlling power, protocols for selecting dynamic frequencies, and procedures for accessing mediums are all components of sharing mechanisms that are very efficient. In order to avoid collisions and preserve spectral efficiency, these methods have been deliberately created. It is common for individuals to make use of cooperative communication and distributed scheduling mechanisms in order to make things more egalitarian and to lessen the amount of noise that is there.

2.5.4 Spectrum Mobility

Altering the wavelength of the emission is the capability of a CR system to easily switch to a different frequency band in the event that a primary user (PU) begins utilising the channel that is already being utilised by another user. This ability is implemented in the event that the PU begins using the channel. It is of the utmost importance that this handoff process be finished as quickly as possible in order to avoid any interference and to keep the lines of communication open. Real-time applications that require continuous connectivity are examples of situations in which the mobility of the spectrum is of the utmost relevance. The cognitive cycle is made of these four functions, which permits the utilisation of the spectrum in a manner that is intelligent, flexible, and efficient in a variety of wireless scenarios. There is the possibility of setting up both a primary user (PU) and a secondary user (SU).

As a result of the use of CR networks, spectrum users can be classified into two basic categories: PUs are licensed users who are the only individuals permitted to utilise a particular frequency band. This type of user is referred to as a PU. The communication that is going place between the main user and themself is not allowed to be interrupted by any other users participating in the conversation. There are many instances, including mobile phone service providers, television services, and satellite systems that have been granted licenses. SU refers to an unlicensed user who makes use of spectrum bands that PUs are not currently utilising. SUs are referred to as SU for short. When a prime user begins using a channel, SUs are obliged to promptly vacate the channel and maintain a strict watch on the spectrum at all times. This will ensure that the spectrum is not compromised in any way. Specific actions are the only ones that SUs is permitted to carry out in order to prevent the production of interference that is damaging. PU and SU work together to create the notion of CR, which is based on the way in which they collaborate with one another. It is necessary to have the capability to precisely detect the activity of the principal user (PU) and to switch between different frequencies in a short amount of time in order to preserve coexistence and maintain compliance with the rules [20], [21].

2.6 SPECTRUM SENSING

2.6.1 Definition

For CR networks, spectrum sensing is a fundamental feature that enables them to locate vacant frequency bands, commonly referred to as spectrum gaps or white spaces, without causing any disruption to licensed PUs. Keeping a constant watch on the spectral environment, determining whether or not primary signals are available, and determining whether or not SUs are able to exploit the channel in an opportunistic manner are the basic objectives of spectrum sensing.

Spectral sensing needs to be extremely accurate in order to guarantee that licensed and unauthorised users do not interfere with one another. In the event that you make a wrong determination that the spectrum is already available, it may result in difficulties for PUs. When you are too cautious when detecting, on the other hand, you run the risk of missing out on the opportunity to make use of the spectrum resources that are available. Spectrum sensing techniques must balance detection probability, false alarm probability, sensing time, and computational complexity, as these parameters directly determine the reliability and efficiency of the sensing scheme [22]. The use of artificial intelligence and cooperative networking frameworks is becoming increasingly tied to spectrum sensing in advanced 5G and future 6G telecommunications systems. This action is being taken in order to improve the reliability of communication in environments that are experiencing high levels of activity and rapid movement [23].

2.6.2 Spectrum Sensing Techniques

To accommodate the requirements of a wide variety of channels, communications, and systems, a large variety of spectrum sensing methods have been developed. This can be accomplished in a variety of ways, each of which varies in terms of the degree of effort involved, the amount of information that must be known in advance, and the degree to which they function well in the presence of any background noise.

2.6.2.1 Energy Detection (ED):

Energy detection (ED), which is also known as radiometry or periodogram-based detection, is the most common way to sense the spectrum in CR networks since it is easy to use, quick to set up, and doesn't require a lot of processing power. It is especially appealing for real-time spectrum sensing uses in 5G and new 6G systems. The fundamental principle of ED is to measure the energy of the received signal over a specified frequency band and compare it to a predefined threshold to determine whether a PU is present.

Key Benefits of ED:

- Easy to set up
- No need to know about PU signals before using them
- Low computational complexity
- Quick decision making that works for real-time systems
- Very good for wideband spectrum sensing

Challenges and Limitations:

- It doesn't work well at low signal-to-noise ratio (SNR)
- Very sensitive to sounds that isn't clear Can't tell the difference between PU signals and noise from other SUs.
- Performance gets worse when the signal fades or shadows.
- In settings with low SNR, the chance of detection goes down a lot, which makes it hard to get reliable spectrum sensing.
- Also, because ED just looks at signal energy, it can't tell the difference between licensed primary signals and noise.

Performance Characteristics:

- Probability of detection (Pd): It is modest to moderate at low SNR.
- Sensing Accuracy: Not very good because it is sensitive to noise

- SNR Requirement: Needs a rather high SNR for reliable detection.
- Real-Time Suitability: Highly suitable due to low processing complexity

ED is still one of the most useful and extensively used ways to sense the spectrum in CR networks. It doesn't operate as well in low-SNR and noise-uncertain conditions, but it's a popular choice for wideband spectrum monitoring in modern wireless systems, including 6G CR frameworks, because it's easy to use, doesn't require a lot of processing power, and can work in real time.

2.6.2.2 Matched Filter Detection

If you already know what the PU waveform looks like, the best technique to sense a spectrum is with matched filter detection (MFD). Matched filtering is the best technique to find signals using classical spectrum sensing methods, especially when the SNR is low. For matched filter detection to work, you need to know what the PU signal looks like (for example, its modulation format, pilot sequence, pulse shape, or frame structure). ED doesn't need this information ahead of time.

Principle of Operation

Matched filtering is based on a linear filter that maximises the output's SNR when there is additive white Gaussian noise (AWGN).

The SU gets a signal that is linked to a known reference signal (template) from the PU. This connection makes the desired signal stronger and cuts down on noise.

The binary hypothesis model is:

$$H0: R[N] = W[N] \text{ (PU ABSENT)}$$

$$H1: R[N] = S[N] + W[N] \text{ (PU PRESENT)}$$

$r[n]$ = signal that was received
s[n] = known PU signal
w[n] = AWGN
Mathematical Model

To acquire the matching filter output, you need to compare the incoming signal to the PU template that you already know:

$$y(n) = \sum_{\{i=0\}}^{\{N-1\}} r(n-i)s(i)$$

where:

s[i] = known template for a PU signal

$r[n]$ = the signal that was gotten

N = the number of samples

y[n] = The output of the matching filter (decision statistic).

In practice, the PU signal is flipped in time and utilised as the impulse response for the matching filter. The method is the same as convolving the received signal with the filter that matches it.

Decision Rule

The decision statistic y[n] is compared to a threshold value λ:

$$y(n) \gtrless_{\{H_0\}}^{\{H_1\}} \lambda$$

- If y[n] > λ, the PU is declared present.
- If y[n] ≤ λ, the spectrum is considered idle.

The threshold is typically determined based on the noise variance and the desired probability of false alarm.

Implementation step:

Here's a short explanation of how the process of matching filters works:

- Signal Preparation
  - Find out about the PU signal ahead of time, such as the type of modulation, the pilot pattern, and the waveform.
  - To build the matching filter, flip the known PU signal in time.
- Convolution and Correlation
  - Use the matching filter to convolve the signal you got.
  - Find the outcome of the correlation.
- Threshold Calculation
  - Find the threshold λ based on how much noise there is and how much detection is needed.
- Decision Making
  - Check the output of the matching filter against the limit.
  - Say if PU is there or not.

Performance Characteristics

- Probability of Detection (Pd)
    - Very high, especially when the SNR is low.
    - Best detection in AWGN conditions
- Sensing Accuracy
    - Very precise when you know the PU signal
    - Better than finding energy
- SNR Requirement
    - Works well even when the SNR is low

Advantages:

Most accurate detection of all standard methods

- Best SNR performance
- Works well in low-SNR situations

Limitations:

- Requires full knowledge of the PU signal
- Is hard to calculate
- Uses more power
- It's challenging to set up in networks with diverse kinds of devices.

Matched filter detection is the best and most accurate approach to feel the spectrum in CR networks when you already know what the main user signal is. It performs better than ED, especially when the SNR is low, but it needs a lot of signal information, which makes the system more complex and uses more power. So, matched filtering works best when the PU's traits are fully known and the environment is controlled. It doesn't perform as well in 6G wireless systems that are often changing and feature a lot of various kinds of devices.

2.6.2.3 Cyclostationary Feature Detection (CFD)

Cyclostationary feature detection (CFD) is a sophisticated method for spectrum sensing that utilises the inherent periodic statistical characteristics of modulated communication signals to identify a PU. Noise that is random doesn't usually fluctuate over time and doesn't have a pattern. Carrier frequencies, modulation techniques, symbol speeds, and cyclic prefixes all cause communication signals to have built-in patterns. Because of these periodic features, the signal's mean and autocorrelation functions change in a cycle over time. This will help you tell the PU signal apart from noise and other signals.

Principle of Operation:

The concept of cyclostationary detection is that:

- Cyclostationary is a property of modulated signals that means their statistical qualities change over time.
- Noise is usually stationary and doesn't have any patterns that repeat over time.

The detector can reliably find the PU even when the SNR is low by looking at these periodic parts.

The major techniques to apply Cyclostationary detection

2.6.2.3.1 Making the Most of Periodicity The method uses the cyclic autocorrelation function (CAF), which looks for patterns in the signal that are received that happen again and again. These repeating patterns are linked to some sections of the PU gearbox, like the carrier frequency and the type of modulation.

The CAF is defined as:

$$R_x^{\alpha}(\tau) = E\left[x(t)\, x^*(t-\tau)\, e^{-j2\pi\alpha t}\right]$$

where:

α = the number of cycles per second

τ = time delay

x(t) = the signal that was transmitted

If $R_x^{\alpha}(\tau) \neq 0$ for some $\alpha \neq 0$, the signal exhibits Cyclostationary behaviour.

2.6.2.3.2 The SCF or Spectral Correlation Function The detection method also includes finding the spectral correlation function (SCF), which is the Fourier transform of the CAF:

$$S_x^{\alpha}(f) = \int_{-\infty}^{\infty} R_x^{\alpha}(\tau)\, e^{-j2\pi f\tau} d\tau$$

The SCF shows which regions of the frequency have the same cyclic frequencies. When the SCF has peaks, there are modulated signals.

2.6.2.3.3 Threshold-Based Decision When the detector figures it out, it checks to see if the SCF is bigger than a certain amount:

- If the SCF is higher than the threshold, PU is there.
- If not, spectrum isn't operating (ideal).

When there is only noise, the threshold is commonly chosen based on how SCF is spread out.

Performance Characteristics

Probability of Detection

- Very high, even when the SNR is low.

Strength

- Very good at blocking out noise that is hard to foresee
- Can tell the difference between signals that vary and noise that stays the same

Signal Identification Capability

- Able to tell the difference between different types of modulation
- Can tell the difference between PU signals and noise

Advantages:

- Works well when the SNR is low
- Is strong against noise uncertainty

- Can tell the difference between different kinds of signals
- Works well when signals cross each other

Limitations:

- Very hard to do using maths
- Needs a long time to watch
- Needs greater strength
- High computational complexity

2.6.2.4 ML-Based Spectrum Sensing

ML-based spectrum sensing as shown in Figure 2.7 is a smart, data-driven way to improve detection in cognitive radio networks (CRNs) by learning about signal qualities from data that has already been seen. ED, matching filter detection, and cyclostationary detection are all examples of traditional spectrum sensing approaches that use predefined statistical models and set thresholds. In contrast, ML-based approaches adjust to changes in the environment by using training and continuous learning systems along with an optimal cooperative spectrum sensing framework in CR networks that enhances detection of weak primary signals.[24], [25], [26].

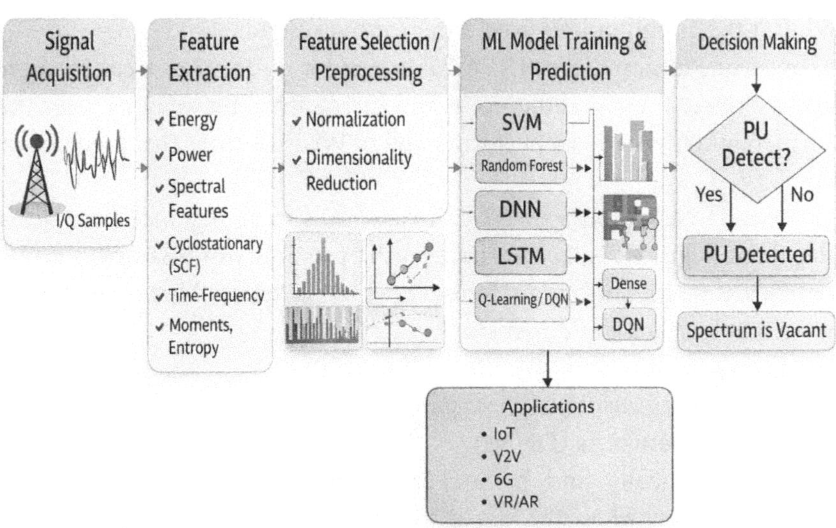

FIGURE 2.7 General framework for machine learning based spectrum sensing.

In dynamic 6G wireless environments characterised by dense deployments, interference, fading, and a low SNR, traditional sensing methods often demonstrate reduced efficacy. Thilina et al. propose cooperative spectrum sensing algorithms for cognitive radio networks based on machine learning techniques used for pattern classification, employing both unsupervised methods (e.g., K-means clustering) and supervised classifiers to distinguish spectrum availability [27]. One of the key advantages of ML-based sensing is its ability to jointly improve detection probability (Pd) while reducing false alarm probability (Pf), thereby enabling more efficient spectrum utilization [28].

2.6.2.4.1 Classification of ML-Based Spectrum Sensing There are a few basic types of spectrum sensing technologies that use ML:

- Classical Supervised Learning
- Deep Learning
- Reinforcement Learning

2.6.2.4.1.1 Traditional Approaches to Supervised Learning Supervised learning approaches use tagged training data to sort spectrum states.

Support Vector Machine (SVM): SVM makes the best hyperplane that distinguishes classes of occupied and unoccupied spectrum as much as feasible. It performs well with tiny datasets and moderate SNR scenarios [29]. But the computations are harder for big systems.

Forest of Random Trees (RF): RF uses a lot of decision trees to determine a choice. It can handle noise and interactions between features that aren't straight lines, thus it performs well in Wi-Fi settings that aren't all the same [30].

k-Nearest Neighbours (k-NN): k-NN is a simple instance-based classifier that assigns class labels based on the proximity of a sample to other examples. It is straightforward to use; however, it has problems with feature spaces that have a lot of dimensions.

2.6.2.4.1.2 Deep Learning Approaches

Deep Neural Networks (DNNs)

DNNs automatically find hierarchical features in raw I/Q or spectral signal data. Even when the signal-to-noise ratio is low and the channel environment is hard to understand, they nevertheless have good detection accuracy [31]. But they need a lot of data to train on and sophisticated processors.

2.6.2.4.1.3 Reinforcement Learning (RL) Approaches Reinforcement learning (RL) helps clever agents learn how to sense and get around on the go. Q-Learning is a model-free reinforcement learning technique that modifies Q-values through incentive feedback. It helps you pick channels that change in CRNs [32].

Deep Q-Networks (DQN)

In dense 6G environments, where there are a lot of possible actions and states, DQN combines deep neural networks with Q-learning to handle them [33–34]. Even when they are strong, convergence stability and computing complexity are still challenges.

TABLE 2.2 Comparison of Machine Learning Techniques for Spectrum Sensing

Technique	Learning Type	Strength	Limitation	6G Suitability
SVM	Supervised	High accuracy with small data	High computation for large data	Moderate
Random Forest	Supervised	Robust to noise	Less interpretable	High
k-NN	Supervised	Simple implementation	Sensitive to parameter k	Moderate
DNN	Deep Learning	Automatic feature extraction	High training cost	Very High
Q-Learning	Reinforcement	Adaptive policy learning	Slow convergence	High
DQN	Deep RL	Handles large state space	High complexity	Very High

TABLE 2.3 Comparison with Traditional Spectrum Sensing Methods The Table- 2.3, discussed the Traditional Spectrum Sensing Methods for 5G and 6G network.

Method	Prior Knowledge	Low-SNR Performance	Complexity	False Alarm Rate
Energy Detection	No	Poor	Low	High
Matched Filter	Yes	Excellent	High	Low
Cyclostationary	Partial	Very Good	Very High	Low
ML-Based	Data-driven	Excellent	Moderate–High	Very Low

CONCLUSION

In the review paper, we discuss the role of cognitive radio in spectrum management for 5G and 6G communication systems. Moreover, the transition from 5G to 6G implies a fundamental paradigm shift in wireless communication, moving from a service-oriented infrastructure to an AI-native, large-dimensional autonomous ecosystem. While 5G effectively introduced the service-based architecture (SBA), network slicing, and initial massive MIMO deployments to support varied industrial use cases, 6G is expected to overcome these barriers by merging space, air, ground, and underwater elements into a single cohesive fabric. Finally, in this review paper, we have also illustrates the AI-Native frameworks, where AI is often treated as an overlay for management, 6G embeds AI directly into the protocol stack (down to the PHY and MAC layers). This enables real-time optimisation, self-healing, and the management of extreme complexity in holographic-ready networks and next-generation radio enablers. Moreover, the synergy between terahertz (THz) communications and ultra-massive MIMO is necessary for obtaining multi-Tb/s peak data speeds. To alleviate the propagation issues of THz frequencies, reconfigurable intelligent surfaces (RIS) will play a crucial role in designing the wireless environment and increasing coverage.

PRACTITIONER'S TAKEAWAY

The proposed the AI-Native frameworks, where AI is often treated as an overlay for management, 6G embeds AI directly into the protocol stack (down to the PHY and MAC layers). This enables real-time optimisation, self-healing, and the management of extreme complexity in holographic-ready networks and next-generation radio enablers. Moreover, the synergy between terahertz (THz) communications and ultra-massive MIMO is necessary for obtaining multi-Tb/s peak data speeds. To alleviate the propagation issues of THz frequencies, reconfigurable intelligent surfaces (RIS) will play a crucial role in designing the wireless environment and increasing coverage.

REFERENCES

1 J. Mitola III and G. Q. Maguire Jr., "Cognitive radio: Making software radios more personal," *IEEE Personal Communications*, vol. 6, no. 4, pp. 13–18, Aug. 1999.

2 S. Haykin, "Cognitive radio: Brain-empowered wireless communications," *IEEE Journal on Selected Areas in Communications*, vol. 23, no. 2, pp. 201–220, Feb. 2005.

3 3GPP TS 23.501, "System architecture for the 5G system (5GS)," *Release* 15, 2018.
4 Federal Communications Commission, "Spectrum policy task force report," ET Docket No. 02-135, Nov. 2002.
5 T. S. Rappaport, et al., "Wireless communications and applications above 100 GHz: Opportunities and challenges for 6G and beyond," *IEEE Access*, vol. 7, pp. 78729–78757, 2019.
6 3GPP TS 38.300, "NR; Overall description; Stage-2," *Release* 15, 2018.
7 E. Dahlman, S. Parkvall, and J. Sköld, *5G NR: The Next Generation Wireless Access Technology*, 2nd ed. Academic Press, Stockholm Sweden, 2020.
8 ITU-R, "IMT Vision – Framework and overall objectives of the future development of IMT for 2020 and beyond," Recommendation ITU-R M.2083-0, 2015.
9 T. S. Rappaport, Y. Xing, G. R. MacCartney Jr., A. F. Molisch, E. Mellios, and J. Zhang, "Overview of millimeter wave communications for fifth-generation (5G) wireless networks," *IEEE Transactions on Antennas and Propagation*, vol. 65, no. 12, pp. 6213–6230, Dec. 2017.
10 M. Shafi *et al.*, "5G: A tutorial overview of standards, trials, challenges, deployment, and practice," *IEEE Journal on Selected Areas in Communications*, vol. 35, no. 6, pp. 1201–1221, June 2017.
11 H. Urkowitz, "Energy detection of unknown deterministic signals," *Proceedings of the IEEE*, vol. 55, no. 4, pp. 523–531, Apr. 1967.
12 A. Sahai and D. Cabric, "Spectrum sensing: Fundamental limits and practical challenges," in *Proceeding of IEEE International Symposium on New Frontiers in Dynamic Spectrum Access Networks (DySPAN)*, 2005, pp. 1–11.
13 Y.-C. Liang, Y. Zeng, E. C. Y. Peh, and A. T. Hoang, "Sensing-throughput tradeoff for cognitive radio networks," *IEEE Transactions on Wireless Communications*, vol. 7, no. 4, pp. 1326–1337, Apr. 2008.
14 I. F. Akyildiz, A. Kak, and S. Nie, "6G and beyond: The future of wireless communications systems," *IEEE Access*, vol. 8, pp. 133995–134030, 2020.
15 W. Saad, M. Bennis, and M. Chen, "A vision of 6G wireless systems: Applications, trends, technologies, and open research problems," *IEEE Network*, vol. 34, no. 3, pp. 134–142, May/Jun. 2020.
16 M. A. McHenry, "NSF spectrum occupancy measurements project summary," in *Shared Spectrum Company Report*, Chicago, IL, USA, Aug. 2005.
17 Q. Zhao and B. M. Sadler, "A survey of dynamic spectrum access," *IEEE Signal Processing Magazine*, vol. 24, no. 3, pp. 79–89, May 2007.
18 Federal Communications Commission (FCC), "Report and order and second further notice of proposed rulemaking: Amendment of the commission's rules with regard to commercial operations in the 3550–3650 MHz band," FCC 15-47, Apr. 2015.

19 D. Cabric, A. Tkachenko, and R. W. Brodersen, "Spectrum sensing measurements of pilot, energy, and collaborative detection," in *Proceeding of IEEE Military Communications Conference (MILCOM)*, Washington, D.C., USA, 2006, pp. 1–7.

20 Y. Zeng and Y.-C. Liang, "Cyclostationary-based spectrum sensing for cognitive radios," *IEEE Transactions on Wireless Communications*, vol. 9, no. 2, pp. 821–830, Feb. 2010.

21 S. M. Kay, *Fundamentals of Statistical Signal Processing, Volume II: Detection Theory*. Upper Saddle River, NJ, USA: Prentice Hall, 1998.

22 T. Yucek and H. Arslan, "A survey of spectrum sensing algorithms for cognitive radio applications," *IEEE Communications Surveys & Tutorials*, vol. 11, no. 1, pp. 116–130, 2009 .

23 H. Sun, A. Nallanathan, C.-X. Wang, and Y. Chen, "Wideband spectrum sensing for cognitive radio networks: A survey," *IEEE Wireless Communications*, vol. 20, no. 2, pp. 74–81, Apr. 2013.

24 Z. Quan, S. Cui, and A. H. Sayed, "Optimal linear cooperation for spectrum sensing in cognitive radio networks," *IEEE Journal of Selected Topics in Signal Processing*, vol. 2, no. 1, pp. 28–40, Feb. 2008.

25 T. Yücek and H. Arslan, "A survey of spectrum sensing algorithms for cognitive radio applications," *IEEE Communications Surveys & Tutorials*, vol. 11, no. 1, pp. 116–130, First Quarter 2009.

26 Y. Zeng, Y.-C. Liang, A. T. Hoang, and R. Zhang, "A review on spectrum sensing for cognitive radio: Challenges and solutions," *EURASIP Journal on Advances in Signal Processing*, vol. 2010, no. 1, pp. 1–15, 2010.

27 K. M. Thilina, K. W. Choi, N. Saquib, and E. Hossain, "Machine learning techniques for cooperative spectrum sensing in cognitive radio networks," *IEEE Journal on Selected Areas in Communications*, vol. 31, no. 11, pp. 2209–2221, 2013.

28 Y. Zhang, Q. Wu, and M. Song, "Energy-efficient cooperative spectrum sensing using machine learning algorithm," *Wireless Communications and Mobile Computing*, 2022.

29 A. Azarfar, J.-F. Frigon, and B. Sansò, "Improving the performance of spectrum sensing using machine learning techniques," in *Proceedings of the IEEE Global Communications Conference (GLOBECOM)*, IEEE, Atlanta, GA, USA 2013, pp. 405–410.

30 F. Restuccia, S. D'Oro, and T. Melodia, "Securing cognitive radio networks: New vulnerabilities and research challenges," *IEEE Communications Magazine*, vol. 54, no. 3, pp. 20–26, Mar. 2016.

31 T. O'Shea and J. Hoydis, "An introduction to deep learning for the physical layer," *IEEE Transactions on Cognitive Communications and Networking*, vol. 3, no. 4, pp. 563–575, Dec. 2017.

32 K. B. Letaief and W. Zhang, "Cooperative communications for cognitive radio networks," *Proceedings of the IEEE*, vol. 97, no. 5, pp. 878–893, May 2009.

33 N. C. Luong, D. T. Hoang, S. Gong, D. Niyato, P. Wang, Y.-C. Liang, and D. I. Kim, "Applications of deep reinforcement learning in communications and networking: A survey," *IEEE Communications Surveys & Tutorials*, vol. 21, no. 4, pp. 3133–3174, Fourth Quarter 2019.

34 Z. Zhang *et al.*, "6G wireless networks: Vision, requirements, architecture, and key technologies," *IEEE Vehicular Technology Magazine*, vol. 14, no. 3, pp. 28–41, July 2019, doi: 10.1109/MVT.2019.2921208.

CHAPTER 3

From Cognitive Radio to Intelligent Spectrum Ecosystems

Foundations, AI-Driven Decision-Making, and Blockchain-Enabled Governance for 5G and beyond

Vinod H. Patil, Sudhir Kadam, Amol Kadam, and Asheesh Tripathi

3.1 INTRODUCTION

Spectrum management is kept in an awkward level. On the one hand, data obtained indicates that every hour, licensed spectrum bands devote huge portions of time lying idle [1]. On the other side, mobile networks are grappling with an exponentially increasing swamp of mobile traffic, and rural consumers in much of the developing world continue to be underserved despite decades of policy work. The discrepancy between these two realities lies, in its essence, in institutional and technological failure a failure to reconcile the way the spectrum is assigned with the way the spectrum is utilised.

This chapter holds the case that the solution to this gap requires more than policy modification. It involves a combination of three technologies which includes artificial intelligence, machine learning, and blockchain, which can all combine to make spectrum management fast enough, intelligent enough, and trustworthy enough to survive into the 5G age and

DOI: 10.1201/9781003731597-3

beyond. Following the path of intellectual development of early cognitive radio studies up to contemporary deep learning and reinforcement learning methods, the chapter explores how blockchain offers the governance infrastructure required by AI-guided spectrum sharing and uses practical testbed experience to actualise these concepts in engineering reality. It ends by mapping the most difficult unsolved problems that the research community should face.

3.2 THE SPECTRUM CHALLENGE IN THE 5G AND 6G ERA

Numbers help here. ITU has estimated data transmission in mobile data to be in hundreds of exabytes monthly before the close of this decade – a number too huge to be intuitively comprehended [3]. What is so tractable is to consider the heterogeneity of applications that generate that traffic: high-definition video, cloud gaming, applications run on lightweight headsets, precision farming sensors reporting moisture in the fields every few minutes, remote robotic surgery with sub-millisecond restrictions on latency, autonomous vehicle coordination with updates every split-second. None of these applications has the same spectrum requirements as any other. And no systematic distribution plan can effectively apply to all in one.

The designers of 5G realised this, at least in part. The 5G New Radio standard brought in what it terms numerology – a parameterisation of the waveform that is flexible enough to allow operators to configure subcarrier spacing and slot duration to the latency and bandwidth profile of the service being provided. The frequencies above 24 GHz were cleared in commercial deployment and provided the operators with raw bandwidth that just did not exist in the 4G spectrum environment. And spectrum sharing mechanisms were designed to be dynamic since the standard was initially established and 5G NR carriers can coexist with legacy LTE on the same frequency bands throughout the transition period [4].

All this did not solve the problem completely though. Practically, it is still in the network planning tools rather than in real time that most of the 5G spectrum management decisions are to be executed. An operator presets the coverage, capacity, and interference limits several weeks or months prior to the launch of service. Narrow changes can be made in real time but not throughout the entire design space. It has not yet become a reality to see the vision of spectrum resources that are dynamically distributed in each place of highest demand, between operators, between services, geographically spread out.

The research on 6G is already heavily ongoing in laboratories in Asia, Europe, and North America, which indicates that the situation is soon going to be significantly more complicated. One of the signature technologies of 6G is called terahertz band communication, and it is defined based on its use on frequencies at which propagation behaviour is so dependent on the environment that even the concept of broad geographic frequency allocation seems a bit ludicrous. Terahertz channels may change in path loss by tens of decibels across a few centimetres depending upon atmospheric composition, temperature, and physical obstructions. Spectrum management in these frequencies must work at centimetre spatial scales and milliseconds timescales, in active monitoring of the physical environment [5].

To this compound the emergence of non-terrestrial networks, the low-earth-orbit satellites arrays being rolled out by SpaceX, Amazon, and One Web on an unprecedented scale, and the problem of spectrum coordination becomes even more difficult. These constellations interfere with earth networks. They continuously vary in their orbital geometries. There was no existing regulatory framework to deal with this, and none of the existing spectrum management technology is up to the task without considerable enhancement by smart and autonomous systems.

The Core Problem

The spectrum challenge in the 5G and 6G era is not simply about having more frequency bands available. It is about having management systems that are fast enough, adaptive enough, and trustworthy enough to allocate whatever spectrum exists to wherever it is most needed, in real time, across the full complexity of modern and future network deployments. That is an AI and governance problem as much as a frequency engineering problem.

3.3 COGNITIVE RADIO: THE FOUNDATIONAL FRAMEWORK

One should take time and study the architecture of cognitive radio not because it answers the problems we have today, but because it puts the problems into perspective. The field unravelled what questions should be asked in the cognitive radio literature, which is about 20 years of work by researchers working quite intensively. It is only against that backdrop that the AI and blockchain solutions that we will discuss later in this paper make sense.

At its most basic, a cognitive radio is a software-defined radio – a radio with physical-layer behaviour represented by software instead of hard-written hardware [2]. This provides it with some leeway which is not available with conventional radios. Its behaviour is governed by the cognitive cycle that follows three steps, namely, observe, reason, act. The radio monitors the electromagnetic world using sensing. It makes logical reasoning regarding what it has seen, considering its needs of service and understanding of the regulatory environment. It can do this by resetting its transmission parameters – frequency, power, modulation scheme, access protocol to exploit available spectral capacity optimally without conflicting with licensed users.

The four functional domains that these three stages project onto adopted the form of the standard organising framework of the cognitive radio literature. The radio detects the spectrum available by spectrum sensing. This is not as easy as it may sound: a poor primary user signal in a fading channel may be easily confused with noise, and a false decision that a channel is free will result in harmful interference. The radio decides what channels to use out of the various channels available to it, and considers such factors as the quality of the received signal, the stability it is likely to maintain, and the needs that the particular traffic it must support demands. Spectrum sharing refers to the protocols according to which several secondary users ensure that they use the same pool of spectrum opportunities without colliding destructively with each other. and spectrum mobility deals with the case where a primary user recovers a previously occupied channel in which a secondary user is already using the channel – the handoff process, which allows the secondary user to proceed with services in a new channel with minimal service interruption.

Hardware Platforms for Cognitive Radio Research: Cognitive radio research has almost co-opted software-defined radio systems based on National Instruments USRP devices and LabVIEW development environment. An example is the USRP 2901, which is 70 MHz to 6 GHz with a 56 MHz peak instantaneous bandwidth and is connected to a host computer via a high-speed USB 3.0 interface [6]. The LabVIEW Communications System Design Suite is a graphical development system enabling researchers to create and experiment with cognitive radio algorithms at real-time including over-the-air experiments, without creating low-level hardware drivers. This combination has enabled reproducible cognitive radio experimentation to be available to academic groups, who would otherwise not have the resources to construct comparable testbeds themselves.

The weaknesses of the classical cognitive radio are now well known. Most commonly used sensing techniques The most common sensing operator used in practice is energy detection, which is sensing that does not need information about the modulation or timing structure of the primary signal, making it fundamentally sensitive to uncertainty due to noise. A single fraction of a decibel difference in the estimate of the noise floor by the detector causes a miscalibration in the detector, and severely impairs its operation. Such detectors as feature-based detectors, which are based on the cyclostationary properties of digital communications signal, work better in noisy conditions but need prior knowledge of the characteristics of the primary signal. Both methods have poor generalisation to the variety of signal types present in real spectrum environments.

Classical cognitive radio is even more limited in decision-making. The majority of them adopt fixed rule tables: when SNR is greater than a certain value, then choose this channel; when interference is greater then choose the other one. These regulations are not capable of evolving to the statistical geometry of the surrounding, are not capable of learning the choices made in the past, and are not capable of predicting the future primary user behaviour. The outcome is a system that makes rationally decision locally, but globally, it is suboptimal.

These are the gaps that are filled by machine learning. It is worth mentioning something on the cognitive radio frame before setting out to delve on how to do it; its long term conceptual value. The four functional domains, sensing, decision, sharing, mobility, and the observe-reason act cycle are still the correct way to structure thinking concerning the management of dynamic spectrum, although the technologies that will realise each of the functions is changing radically. Good structures outlive the definite technologies that embodiment them.

3.4 ARTIFICIAL INTELLIGENCE AND MACHINE LEARNING FOR INTELLIGENT SPECTRUM MANAGEMENT

This last ten years has witnessed the evolution of machine learning as a potentially advantageous theoretical tool in the communications literature to a critical element of a realistic system design. This change has not been equally spread at all issues, some uses have been greatly advantaged more than others, yet in the field of dynamic spectrum management, the benefits that ML can bring are enormous and well-reported. The next thing is to look at the way that machine learning changes each of the four fundamental operations of the cognitive radio.

3.4.1 Deep Learning for Spectrum Sensing

Turning spectrum sensing into a mathematical skeleton, that is a hypothesis test. Based on a received signal, determine whether a primary user is present or not. That framing immediately gives one an idea of the difficulty with which classical statistical detectors can deal with it: the optimal statistic of the test is sensitive to both signal and noise distributions, and in real channels, neither these distributions is known nor stationary. Deep learning does not have this issue, or at any rate it does not have it in the same manner. A convolutional neural net that is trained on spectrograms, or images of received signals (frequency bands) learns to discriminate the visual patterns of primary user activity without being informed about what these patterns are [7]. Data informs the network of what occupied spectrum looks like and what vacant spectrum looks like. This is a considerably different epistemology to the threshold-based detector, and results in a considerably different performance profile. At low SNR, the CNNs that have been trained on realistic channel data have demonstrated several decibel improvements in detection over energy detection which directly converts into more reliable sensing in the conditions at the edges where mis-detection is the most significant.

Trends of the primary user activity are also important. A TV message takes up the channel it is meant to occupy for hours. The uplink traffic of a cellular base station peaks during the commuting time and falls off at night. A weather radar is swept in a rotational predictable pattern. None of this time structure is used by a detector that uses each sensing observation separately. These patterns can be learnt using long-term memory networks designed to capture long-range dependencies in sequential data and used to predict the location of spectrum opportunities even before they occur in reality to facilitate what researchers refer to as proactive sensing [8]. A proactive system does not monitor all of the bands at any given time but focuses its sensing effort on those bands that are most likely to become available or those that are most likely to be reclaimed soon.

Several distributed nodes can collaborate in sensing to make a shared decision in the concept of cooperative sensing, which has been studied since the early days of cognitive radio. The rationale is obvious, no individual node can see all primary transmitters, but a node network can offer significantly greater coverage. The difficulty is aggregation: how can one take noisy, possibly conflicting reports of heterogeneous nodes to an accurate group decision? One solution is provided by deep learning, where the learned fusion architecture considers the contribution of nodes in direct proportion to their perceived reliability instead of considering all reports equally reliable [9]. However, it is also the source of a security issue – a bad or malicious node that introduces a false report has the potential to corrupt the collective decision, should fusion mechanism not be adversarially robust. It is a research issue under active investigation, and the problem of spectrum decision is directly related to the blockchain governance mechanisms that are described in Section 5.

3.4.2 Reinforcement Learning for Spectrum Decision and Optimisation

Spectrum decision. Now you are required to make a choice, understanding that this choice will influence the possible choices in the future, and it is queued that the environment, namely primary user behaviour, interference levels, and channel quality are in some degree unpredictable. Classical methods of optimisation cannot solve this type of problem since they need an open and total model of the environment, and you will never have it. Reinforcement learning is successful since it does not involve the need of reinforcement.

In RL, an agent is exposed to the environment to learn the policy of making decisions. It attempts actions, checks the consequences, gets feedbacks of what can be considered as a good consequence, and slowly modifies its policy to prefer those actions that yield increasing rewards with time [10]. An agent in the spectrum context may get a reward, high, when selecting a channel with good quality and no primary user interference and negative when selecting a channel which proves to be occupied or causes interference. Throughout thousands of trainings, the agent becomes convinced of the channels that are likely to yield good results, under which conditions within the environment.

An early and extremely popular implementation of this concept to spectrum management was Deep Q-Networks, which are deep neural networks that approximate the optimal action-value function that endowed discrete state spaces cannot represent due to large state spaces. An agent that is trained using DQN can discover how to avoid the channels in which primary activity is predictable, and to exploit quality disparities across channels and do a balance between exploring unfamiliar channels and exploiting known-good channels [10]. More recent designs such as proximal policy optimisation and soft actor-critic, operate continuous action spaces with more grace, which is important when the decision needs to involve not only channel selection but also power level and modulation scheme. Table 3.1 summarises these reinforcement learning approaches, comparing their key strengths, main limitations, and best-fit applications within cognitive radio spectrum management.

Multi-agent reinforcement learning applies such in multiple ways to networks where many secondary users must make a joint spectrum control decision in the absence of an overall coordinator. This is very crucial in actual deployments. Centralised optimisation may be effective in

TABLE 3.1 Comparative Overview of Reinforcement Learning Approaches for Cognitive Radio Spectrum Management

RL Approach	Key Strength	Main Limitation	Best Fit in CR	Reference
Deep Q-Network (DQN)	Handles discrete action spaces well; widely validated	Slow to converge; large state spaces problematic	Channel selection in multi-band deployments	Mnih et al. [10]
Proximal Policy Optimisation (PPO)	Stable training; naturally handles continuous actions	Computationally intensive	Joint power/channel/modulation optimisation	Schulman et al. [11]
Multi-Agent RL (MARL)	Decentralised; scales to many users without central control	Non-stationary learning environment complicates convergence	Distributed CR; 5G/6G network slicing	Zhang et al. [12]
Model-Based RL	Sample-efficient; integrates prior domain knowledge	Model mismatch propagates into policy errors	Proactive handoff using traffic prediction models	Moerland et al. [13]

a controlled experimental environment, but the real networks consist of rival operators, devices of heterogeneity, and dynamic topologies that cannot be monitored by some central controller. In MARL, every device trains a policy, but considers other devices as environmental factors. Having the right reward design – rewards that are used to reconcile individual incentives to system-level efficiency – then these individually selfish learners can converge on co-operative solutions that can nearly match the performance of centralised optimisation without having to be centralised [12]. It is a classy product with a significant practical value.

3.4.3 Federated Learning: Intelligence Without Centralised Data

There is a very conspicuous conflict in the application of machine learning in spectrum management that does not necessarily receive the attention that it warrants. The proper training of a decent deep learning model will need data, and lots of it, and hopefully data that is representative of the entire variety of the environments that the model will find application. However, a network of many independent devices provides a thin slice of that diversity to its individual devices. The apparent answer the most logical thing to do to this is centralise the spectrum monitoring of all users in one server and train it by putting pressure on that server, but this is exactly the failure point since it forms an entry of communication congestion, poses privacy risks as far as location and activity information are concerned, and creates a point of failure. One way out is federated learning, which was presented by McMahan and others when it came to keyboard prediction on mobile phones [14]. In this model, the devices learn a local model update using their data. It then transmits that update a gradient or a compressed weight vector, but not the actual data.

The server adds together updates of numerous devices (usually by some form of weighted averaging), generates a better global model, and transmits it back to the next process of local training. Raw observations do not ever quit the device. When applied to spectrum sensing, federated learning can enable the community of cognitive radio nodes to collectively design a sensing model which captures the complete diversity of their radio environments, with no device disclosing its raw spectrum measurements to any other party. Recent experimental studies indicate that federated sensing models can achieve the same detection performance as federated models trained on centralised data, yet with much less uplink communication bandwidth, which is a significant practical benefit in bandwidth-constrained network environments [15].

A number of complications are still worth mentioning. The local data distribution between various devices do not follow an identical distribution – a device in charge in a dense urban centre will see a very different spectrum environment compared to a device in a rural area. This non-IID structure may lead the global model to a poor solution on certain local environments. To some extent, personalised federated learning methods, that enable each device to optimise the global model on its data following aggregation, solve this but at the expense of increased complexity. Security issues are also at stake: a rogue device can provide poisoned gradient updates that are meant to deteriorate the performance of the global model. This risk can be curtailed with secure aggregation protocols dependent on different privacy, and it is in this direction that – as noted in Section 5 – blockchain-based mechanisms of validating model update provenance are under investigation. Classical cognitive radio is essentially responsive. It identifies an issue – a main user on the existing channel, a loss of signal quality – and reacts to it. The response takes time. In the process, there is interference, poor service quality and low performance of the system.

3.4.4 Proactive Spectrum Management: From Reaction to Anticipation

Machine learning allows an alternative strategy. Since ML models are capable of capturing temporal trends in spectrum behaviour, then they are able to predict where issues will occur even before they occur. Occupancy predictors based on LSTMs trained on past channel usage data have also been able to make meaningful predictions of primary user activity with tens of milliseconds to several minutes of horizon length depending on the regularity of the primary service traffic pattern [8]. Such a predictor available to a secondary user can initiate the search of a replacement channel prior to its reuse by the primary user, schedule its most latency-sensitive transmissions when the primary user is predicted to be busy, and not attempt to begin long transmissions on channels which it suspects will be preempted by the primary user.

The identical proactive reasoning can be applied to interference control, predictive network slice allocation traffic load prediction, predictive beamforming in millimetre-wave systems, where a beam alignment error, such as a user moving behind a barrier, etc., is a practical loss of spectrum access, which can be avoided by proactive tracking. The principle involved is the same in all of them: not to wait till the environment drives a change, but to take advantage of your knowledge about the inclinations of the environment, and get one step ahead of it.

3.5 BLOCKCHAIN TECHNOLOGY: BUILDING TRUST INTO SPECTRUM GOVERNANCE

Sensing and decision-making excellence is only half the solution to spectrum sharing problem. The second aspect – which is arguably the more difficult aspect after leaving the lab – is governance. In a dynamic spectrum sharing between possible parties that might not know one another, might actively interfere with other parties, and may be subject to different national regulatory frameworks, the issue of trust becomes inevitable. Who maintains the record of authority over who has been assigned what spectrum, when and on what conditions? Who verifies compliance? Who arbitrates disputes? And at what speed do these processes occur when even the allocations are being made in milliseconds?

One solution has been suggested as centralised spectrum brokers: trusted third parties that add records of allocations and enforce policies: such solutions have been proposed in the Spectrum Access System that regulates Citizens Broadband Radio Service (CBRS) use in the United States, and practical solutions exist, such as the Spectrum Access System. But there are problems of centralised brokers. They present single points of failure. They demand that there is confidence in the broker by all parties on the neutrality and competence. They have a low rate of transaction processing. When deploying to a multi-national, no single party may be able to be recognised as authoritative by all parties.

Another architecture of the trust problem is the blockchain technology. The key point here is that a distributed registry kept by a group of peers, where each successive record is cryptographically chained to all the previous records, could serve as a source of common truth that is resistant to tampering, and which does not need any trusted party. Blockchain records are, practically, unwritable: to modify any historical record would require recomputation of the cryptographic proof-of-work (or equivalent) on all the later records, and this is computationally infeasible in a well-designed network. And smarter contracts, self-executable programs that run on the blockchain, can be used to automatically enforce the rules of accessing the spectrum, computing compensation, and initiating dispute resolution processes, without human involvement [16].

3.5.1 Choosing the Right Blockchain Architecture

The most famous and technically developed public blockchains, such as Ethereum, are not a good fit to spectrum management in their present form. Even with the switch to a proof-of-stake consensus, the main chain can only handle tens of transactions per second, and this capacity has not yet changed following a switch to the PoS consensus. The latency of transactions, although lower than previous models, remains in seconds or tens of seconds – much too slow to be used in real-time spectrum access. And gas costs add a cost burden to make micro-transactions, which provide access to a channel a matter of a few hundred milliseconds, economically ineffective. All these limitations are solved in permissioned or consortium blockchains, with the most common example being deployed in enterprise environments, Hyperledger Fabric [17].

A consortium blockchain may only be accessible to a limited number of practiced, verified group of organisations in the spectrum case, this could be a group of licensed operators, a national regulator, and a group of large vertical industry consumers. The mechanism of consensus, rather than asking the participants to perform proof-of-work or proof-of-stake calculations and remain anonymous, is based on a set of identifiable and responsible parties called the endorsing peers. This enables significantly greater transaction throughput (in optimised setups, hyperledger fabric can support thousands of transactions per second), reduce and more predictable latency, and support custom privacy policies that permit sensitive commercial data to be selectively shared with all participants instead of being shared with everyone.

In majority of spectrum management applications of the 5G era, consortium blockchain architectures are the appropriate initial point of departure. They offer the guarantees of trust and immutability which are important, without the cost of performance that would render public blockchains impractical. The difference between the two approaches can be reduced as the technology continues to evolve and more layer-2 scaling solutions to public blockchains become available.

3.5.2 Smart Contracts for Automated Spectrum Access

The most obvious short-term way to use blockchain to manage spectrum is automated (smart contract-based) spectrum access lifecycle. Conceptualise a smart contract of spectrum access as an enforceable machine read lease agreement. Such terms as frequency bands, geographic area, maximum power level, time window, compensation rate per unit of access time, conditions under which termination should be triggered, are coded into the contract and sent to the blockchain. When deployed, neither can a party make unilateral changes to the terms and the immutability of the blockchain ensures this. Any request by a secondary user is authenticated by the contract and a grant is registered on the blockchain, the secondary user is notified to start transmitting and the relevant compensation is transferred to the primary license holder in a transaction that can be completed in just under a second on a consortium blockchain. Both the parties do not have to trust each other more than the logic of the contract demands – it is the whole point [16]. More detailed access policies are also easily achieved using smart contracts.

Contract logic can be used to implement dynamic pricing, where the amount of compensation varies according to the current demand, the time of the day, or the available capacity, and would not need to be renegotiated manually. Interference protection clauses, in which, when the transmissions of a secondary user are found to create harmful interference with primary user transmissions, the secondary user is automatically punished, suspended and recorded, can be incorporated directly in the contract. This changes the costly and slow regulatory implementation, which was in place, to an automatic and auditable mechanical implementation.

3.5.3 Spectrum Auctions and Tokenised Rights: A New Market Model

Going even further into the future, blockchain can permit radically different market forms of spectrum allocation that extend far beyond the automation of bilateral contracts by smart contracts. Conventional spectrum auctions – run by national regulators, generating long-period exclusive licences at coarse geographic granularity were developed in an age when spectrum administration occurred at human decision rates. The sums of money that were involved and the duration of time covered made the processes of regulations and enforcement of the law feasible. However, with shorter spectrum allocations, those of smaller geographic scope and more dynamic nature, the overhead of the conventional auction mechanism becomes prohibitive.

Spectrum auctions based on blockchains – where bidding, allocation and settlement are executed as smart contracts on a distributed register – have the potential to lower the overhead of spectrum trading by orders of magnitude [18]. The auction regulations are also coded in a contract open to everyone involved in the auction, which excludes manipulation by the auctioneer. The settlement is automatic and almost instant. The complete history of all bids and allocations is now recorded on the blockchain and available to regulators, which is more transparent to post-auction processes than traditional processes usually can be.

Another facet is the idea of tokenised spectrum rights, where rights are represented as non-fungible tokens on a blockchain: this allows rights to be further fragmented, to be traded, and to be reassembled in secondary markets just as easily as any other blockchain asset [19]. A major licensee with surplus capacity in a given area would have the ability to sell granular access rights, in an automated market, without necessarily having to negotiate with each potential secondary user individually. It is a truly novel market structure and although considerable regulatory effort must first take place before that can run on scale, spectrum regulators in the United States, UK, and the EU have started early scouting of the concept.

3.6 CONVERGING AI AND BLOCKCHAIN: A UNIFIED SPECTRUM ARCHITECTURE

AI and blockchain are commonly referred to as distinct technology paths, both having their research community, their industry supporters, and their implementation path. That is a fallacy in the context of spectrum management. The two technologies will deal with the complementary aspects of the same basic problem, and the best solutions will be grounded on the integration of both.

This is the very centre of the argument. This is speed and optimisation offered by AI and machine learning. They can feel the spectrum, know its future and make allocation decisions within the range of time between micro-seconds and milli-seconds which dynamic access demands. However, AI models are non-transparent, their decisions are difficult to examine, and where AI agents are taking decisions with financial and regulatory impacts, the model decided so cannot be used as justification to justify a disputed allocation. This is where blockchain enters the picture. Blockchain offers transparency and immutability and automated enforcement. It lacks the capability to arrive at decisions quickly, but it can render decisions – and the logs of decisions undertaken by AI systems – responsible, verifiable, and automatically binding. One cannot do without the other layer.

Three logical layers make up an architectural framework of an integrated intelligent spectrum management. First is the perception layer: a distributed network of SDR-enabled devices that do spectrum sensing and report their observations to their higher layers. The intelligence layer is in the middle: occupancy prediction, interference detection, and channel quality prediction ML models that are trained with federated learning and deployed at edge nodes; RL agents that decide channel selection and power control and deep learning classifiers that identify signal and modulation. The top is the layer of governance: a consortium blockchain system operating smart contracts that capture allocation events, compensate, store interference reports, and keep the shared audit trail which can be independently checked by regulators, operators and secondary users [20].

These layers are in a two-way and continuous interaction. Recommendations on allocating resources to enhance the intelligence layer are generated and implemented by the governance layer. The patterns of access, compliance events, and log of interference of the governance layer are training data and regulatory feedback of the models of the intelligence layer. In the long term, this feedback loop implies that the system will become informed about its governance history, the more the system allocates, the better decisions are made depending on which patterns of access have created interference, disputes, or under-utilisation in the past.

There is one credible issue with this architecture that should be discussed with honesty: the latency between the AI and blockchain layers is not matching. Decisions in spectrum access must be made within milliseconds; blockchain confirmation of transactions can take hundreds of milliseconds to seconds even with a fast consortium ledger. The solution to this is two tier commit model. The AI agent decides and determines its access locally and on-the-fly. The respective blockchain record is asynchronously recorded. Periodic reconciliation against local records and blockchain records is done to confirm that records are equivalent and a mismatch between the records results in automated dispute resolution. This leads to the fact that the AI layer is faster and will determine the access latency, whereas the blockchain guarantees are faster and will determine the accountability and auditability [21]. You lose real time blockchain finality, and in return, you obtain feasible performance and regulatory transparency is not compromised.

Integration Principle

In a well-designed intelligent spectrum system, AI provides the speed to make the right decision at the right millisecond. Blockchain provides the trust to prove that the right decision was made and to automatically settle the obligations it creates. Designing either layer without the other produces a system that is either fast but ungovernable or transparent but too slow. The value is in the integration.

3.7 WHAT THE TESTBED TEACHES: GROUNDING THEORY IN ENGINEERING REALITY

Something that cannot be entirely understood through theoretical analysis and simulation is the way actual hardware demands constraints of its own, the timing properties, and modes of failure of systems that appear to be perfectly clean on paper. The experience developing and running a real-time cognitive radio testbed based on NI USRP 2901 hardware and LabVIEW Communications Suite, with a 70 MHz to 6 GHz frequency span and VERT 900 and VERT 2450 omnidirectional antennas, has offered the author some insights that are worth sharing expressly, not that they highlight the theoretical picture, but because they contribute to it in ways that are relevant to real deployments.

The former is regarding measures of signal quality. The concept of channel quality in simulation is almost always represented by a single figure: signal-to-noise ratio. The channel that you use is the one that has the best SNR. Practically, SNR only gives half the story. Total Harmonic Distortion – the energy which a transmitter inevitably emits at the multiples of the fundamental frequency (which was typically comparable to the frequency of the consumer electronics nearby) – was always observable in testbed measurements at 5 GHz, especially when neighbouring consumer electronics were on. Interference caused by THD at harmonics may cause impairment of the reception of other channels even though the channel being used in the transmission appears clean in terms of SNR. Equally, signal-to-noise and distortion ratio which quantifies the loss of signal linearity characterises why higher order modulation mechanisms such as 64-QAM are indeed practical at a certain link margin. An algorithm to select a channel based on the SNR only, and disregarding these effects, will always overrate the throughput that is possible, and can pick channels that, in the real world, will do worse than others it discarded. The moral of the ML-based decision systems: the feature set to the model is as significant as the model architecture. Multi-metric channel quality representations achieve much higher SNR baseline levels.

The second lesson is that of integration latency. The cognitive radio research papers nearly always test sensing, decision, sharing, and mobility functionalities individually, with experiments that are specially optimised to test each of the functionalities. The empirically surprising aspect of the integration latency is manifested when you simultaneous run all the four functions in a single real-time system. Switching between receive (sensing) mode and transmit mode on the USRP platform creates a hardware settling time which, unless explicitly considered in the duty cycle management, provides a detection window through which primary user signals may not be visible. The broader this window, the greater the number of primary user arrivals missed by the system, and the greater the damage caused by interference. The sensing-transmission duty cycle is not a minor implementation consideration; it is a first-class system design issue which becomes even more urgent when sensing is required to span broader bandwidths and more rapid primary user behaviour.

The third observation is institutional and not technical but it can be the most practically important of the three. The spectrum sharing protocol in the testbed was a first come first served sharing that was implemented and a database maintained centrally that registered which secondary users could use which channels at any point in time. This was precisely as intended within the controlled conditions found within the laboratory environment of a single research group. However, it was already clear that any field implementation concerning independent organisations would result in an issue of trust that the protocol fails to address. What does secondary user A know about the allocation database being in the real state of the allocations and that the secondary user B has not been granted access to the same channel? How does the primary licence holder know that the second user who was allowed access really used access time that was no more than that time? The adjudication of claims of interference is how? No answers to any of these questions can be found in a centralised database architecture, and the testbed brought their immediate need home to them. This is not a critique of the testbed design, and it was designed to measure RF performance, rather than governance, but it is an example, on first principles, of why the blockchain governance layer that is described in Section 5 is not an academic embellishment. It is a functional requirement.

3.8 OPEN CHALLENGES AND THE ROAD AHEAD

It is as crucial to recognise what we are yet unaware of as to map what we are. The spectrum management vision outlined in this chapter of an integrated AI-blockchain is sound and technically based, yet the difference between the most current research demonstrations and the production implementation in an actual 5G network is still significant. These are the areas where it is the most difficult to bridge.

3.8.1 Security and Adversarial Robustness

The AI and the blockchain elements of an intelligent spectrum system both offer security attack surfaces that are not yet sufficiently hardened. Adversarial inputs are susceptible to the deep learning models on the AI side, which is a carefully designed signal perturbation that, although nearly imperceptible by humans, causes the neural networks to draw disastrously incorrect classifications [22]. The adversary in the spectrum sensing scenario is able to potentially compromise a sensing model by injecting an adversarially designed signal in the environment to make it either classify an occupied channel as empty (allowing the attacker to remarkably operate without being detected) or an empty channel as occupied (denying secondary users access to which they are entitled to have). One of the biggest open problems is building sensing architectures that are certifiably robust to adversarial perturbations, that is, they are empirically robust in specific attack scenarios, but are provably robust against mathematically defined threat models.

Smart contracts are unalterable on the blockchain side once they are placed in the blockchain. This, broadly speaking, is an attribute – it, that is, does not permit any of the parties to change the terms of an agreement unilaterally. However, when a bug or exploitable design flaw is found in the contract it is a very heavy burden. Hundreds of millions of dollars have been lost in a number of high-profile breaches of contracts in the financial blockchain ecosystem. A defective contract in a spectrum management setting might improperly provide secondary access to licensed bands, fail to impose penalty in case of interference or enable recovery without access privileges. The correct mitigation is formal verification of contract logic, i.e. mathematically demonstrating that the contract is correct when subjected to all possible inputs, but is technically challenging and commercially unusual. It should be turned into a standard practice.

3.8.2 Standardisation and Regulatory Frameworks

Technical solutions that are not able to cross operator, vendor, and national borders will not achieve their potential. The current state of the AI-driven and blockchain-enabled spectrum management standards is very fragmented, which may easily lead to considerable delays in the deployment unless mitigated by a cohesive algorithm of AI-driven spectrum sharing across operator borders. 3GPP Release 18 specification is the first to introduce meaningful intelligent network automation capabilities in 5G Advanced, but a standardised method of AI-driven spectrum sharing across operator boundaries has not yet been standardised. The Experiential Networked Intelligence group of ETSI and the IEEE torque of the Dynamic Spectrum Alliance are working on the respective frameworks but convergence between these two parallel projects will require years to materialise [23]. The technology is even farther ahead of regulatory structures.

The CBRS of the United States and the shared access licensing regime of Ofcom of the UK are helpful precursors of managed dynamic sharing, though neither was intended to support AI-driven decisions or blockchain governance. It will be necessary to continue engaging regulators to extend those frameworks or create new ones to allow the application of machine-speed allocation decisions, the enforcement of smart contracts by automated methods, and blockchain-based audit trails, which is not offered by most technology developers at this time. Energy efficiency and sustainability are more important elements in construction that should not be overlooked or given insufficient attention.

3.8.3 Energy Efficiency and Sustainability

This difficulty is given less attention than it deserves. The inference of AI requirements to all sensing intervals and blockchain transactions to all spectrum access events creates a computational overhead that is multiplied by the number of devices in millions. Network-edge deep learning inference is cheaper by orders of magnitude than network training; however, it is never free, and in the case of a national 5G network, the aggregate energy cost can be large. Permissioned ledger blockchain consensus is more power efficient than proof of work mining, and still has overheads, which must be offset against the value of governance delivered. It must research the intersection of green AI, energy-efficient signal processing, sustainable distributed ledger design, and this is an opportunity to differentiate vendors ready to invest in it.

3.8.4 Terahertz and Non-Terrestrial Networks: The 6G Frontier

The requirements of spectrum management that the 6G research agenda places before current AI and blockchain frameworks are not made to support, and that it is truly difficult to scale it to those requirements. A spectrum management decision at terahertz frequencies must be able to run at centimetre scale spatially and milliseconds scale in time, in real-time following the physical environment. The ML models required to do so should be lightweight so as to execute on edge devices with stringent power constraints, and they should be able to operate on sensor data types (terahertz channel measurements) where large training datasets are yet to be available, and they need to make decisions faster than demonstrated by any current RL deployment. In non-terrestrial networks the orbital geometry is a continuous change, the propagation environment is radically different than in terrestrial deployments and coordination across national jurisdictions and regulatory regimes is a complexity to governance that blockchain frameworks have not been proven to handle. These are open research issues [5].

3.9 CONCLUSION

The thesis of this chapter can be summed up in a simple manner. Spectrum is not limited in the administrative sense only – it is currently under-exploited due to the fact that the systems governing its utilisation cannot match the dynamism of contemporary communications. To make those systems capable machine intelligence is needed to manage speed and complexity, and distributed trust infrastructure is needed to manage governance at scale. Cognitive radio provided us with the theoretical lexicon. The rule-based foundations of cognitive radio could never support sensing and decision capabilities, which AI and machine learning are now offering. The trust layer that is commercially and regulatorily viable with multi-party spectrum sharing is being provided by blockchain.

The most interesting thing to me with this convergence, having been in this field some years and having pondered long and hard over where this convergence was going when editing this volume is that this convergence is not merely a technical extension to the spectrum management toolkit. It is a redefinition of the concept of spectrum management. During the fixed-allocation period, it was more-or-less of a system of property rights: you bought a license, you owned a band, you did whatever you pleased with it within wide limits. During the cognitive radio it became an engineering problem: make radios intelligent enough to locate and use dynamically unused spectrum. In the economic institution this book is describing, it is a market structure, maintained by AI and blockchain, where spectrum access is continuously directed to its high value uses, under transparent regulations that safeguard the public interest and automatically self-enforce.

That is an ambitious vision. The succeeding chapters supply the technicality that is required to comprehend how ambitious it is and how accomplishable it is. This first chapter has made an attempt to give the map; the land is there.

PRACTITIONERS TAKEAWAYS

Spectrum scarcity is, at its heart, a governance failure rather than a shortage of radio frequencies. Smarter utilisation through dynamic access often delivers more capacity than new physical infrastructure ever could and practitioners need to start treating it that way.

The cognitive radio framework still holds. Its four functions sensing, decision, sharing, and mobility remain the right blueprint, but the tools implementing each one need to be AI-driven now, not rule-based. Deep learning for sensing and reinforcement learning for channel decisions are mature enough for real deployments; federated learning is close behind. Getting the reward function right in RL systems is non-negotiable optimise selfishly and you'll cause more interference than you solve.

On the blockchain side, skip public Ethereum and go straight to permissioned ledgers like Hyperledger Fabric. Smart contracts must anticipate disputes primary user return, interference disagreements, payment failures before they arise, not after. Blockchain audit trails are a genuine regulatory asset, not just a technical detail.

Integration is where it gets hard. The two-tier commit model handles the latency gap between AI speed and blockchain settlement. Build it in from the start.

Security, energy efficiency, and dataset generation for terahertz bands are all underinvested. That is where the next wave of work needs to go.

REFERENCES

[1] Federal Communications Commission, "Spectrum policy task force report," ET Docket No. 02-135, Nov. 2002.

[2] J. Mitola III and G. Q. Maguire Jr., "Cognitive radio: Making software radios more personal," *IEEE Personal Communications*, vol. 6, no. 4, pp. 13–18, Aug. 1999.

[3] International Telecommunication Union, "IMT traffic estimates for the years 2020 to 2030," ITU-R Report M.2370-0, Geneva, Jul. 2015.

[4] 3GPP Technical Specification 38.300, "NR; NR and NG-RAN overall description," Release 17, 2022.

[5] F. Tariq, et al., "A speculative study on 6G," *IEEE Wireless Communications*, vol. 27, no. 4, pp. 118–125, Aug. 2020.

[6] National Instruments, "NIUSRP-2901 software defined radio device user manual," Austin, TX, USA, 2018.

[7] T. J. O'Shea and J. Hoydis, "An introduction to deep learning for the physical layer," *IEEE Transactions on Cognitive Communications and Networking*, vol. 3, no. 4, pp. 563–575, Dec. 2017.

[8] X. Xing, T. Jing, W. Cheng, Y. Huo, and X. Cheng, "Spectrum prediction in cognitive radio networks," *IEEE Wireless Communications*, vol. 20, no. 2, pp. 90–96, Apr. 2013.

[9] Z. Chen, N. Guo, Z. Hu, and R. C. Qiu, "Experimental system for cognitive radio based on deep reinforcement learning," *IEEE Access*, vol. 9, pp. 25614–25629, 2021.

[10] V. Mnih, et al., "Human-level control through deep reinforcement learning," *Nature*, vol. 518, no. 7540, pp. 529–533, Feb. 2015.

[11] J. Schulman, F. Wolski, P. Dhariwal, A. Radford, and O. Klimov, "Proximal policy optimization algorithms," arXiv:1707.06347, Jul. 2017.

[12] K. Zhang, Z. Yang, and T. Başar, "Multi-agent reinforcement learning: A selective overview of theories and algorithms," in *Handbook of Reinforcement Learning and Control*, Cham, Switzerland, Springer, 2021, pp. 321–381.

[13] T. M. Moerland, J. Broekens, A. Plaat, and C. M. Jonker, "Model-based reinforcement learning: A survey," *Foundations and Trends in Machine Learning*, vol. 16, no. 1, pp. 1–118, 2023.

[14] B. McMahan, E. Moore, D. Ramage, S. Hampson, and B. A. y Arcas, "Communication-efficient learning of deep networks from decentralized data," *Proceedings of 20th AISTATS*, vol. 54, pp. 1273–1282, Apr. 2017.

[15] Y. Zhao, et al., "Federated learning-based spectrum sensing for cognitive radio networks," *IEEE Communications Letters*, vol. 25, no. 5, pp. 1631–1635, May 2021.

[16] M. Weiss, S. Egelman, and P. Bhatt, "Blockchain for spectrum sharing: Use cases and conceptual framework," in *Proceedings of the IEEE DySPAN*, Seoul, Korea, IEEE, 2018, pp. 1–8.

[17] E. Androulaki, et al., "Hyperledger Fabric: A distributed operating system for permissioned blockchains," in *Proceedings of the 13th ACM European Conference on Computer Systems*, Porto, Portugal, ACM, Apr. 2018, Art. no. 30.

[18] H. Wang et al., "Negotiable auction based on mixed graph: A novel spectrum sharing framework," *IEEE Transactions on Cognitive Communications and Networking*, vol. 3, no. 3, pp. 390–403, Sep. 2017.

[19] Federal Communications Commission, "Tokenized Spectrum Rights and Secondary Markets: An Exploratory Discussion," FCC White Paper, Washington DC, 2023.

[20] M. B. Weiss, K. Werbach, D. C. Sicker, and C. E. Caicedo Bastidas, "On the application of blockchains to spectrum management," *IEEE Transactions on Cognitive Communications and Networking*, vol. 5, no. 2, pp. 193–205, Jun. 2019.

[21] A. Jindal, V. Aujla, and N. Kumar, "Blockchain-based secure data storage and resource management for distributed cognitive radio networks," *IEEE Transactions on Industrial Informatics*, vol. 16, no. 6, pp. 3929–3939, Jun. 2020.

[22] B. Flowers, R. M. Buehrer, and W. C. Headley, "Interpretable deep learning for blind modulation classification," in *Proceedings of the IEEE MILCOM*, Los Angeles, CA, IEEE, Oct. 2019, pp. 1–6.

[23] ETSI GS ENI 001, "*Experiential Networked Intelligence (ENI); ENI Requirements*," vol. 3.1.1, Sophia Antipolis, France, European Telecommunications Standards Institute, 2022.

[24] V. H. Patil and S. K. Oza, "A testbed design of spectrum management in cognitive radio network using NI USRP and LabVIEW," *International Journal of Recent Technology and Engineering*, vol. 8, no. 2S8, pp. 1–8, Aug. 2019.

[25] I. F. Akyildiz, W.-Y. Lee, M. C. Vuran, and S. Mohanty, "Next generation/dynamic spectrum access/cognitive radio wireless networks: A survey," *Computer Networks*, vol. 50, no. 13, pp. 2127–2159, Sep. 2006.

CHAPTER 4

Reimagining Spectrum Management

Foundations for AI-Driven 6G Networks

Kakade Sandeep, Nitin Shivale, Shrishail Patil, Gayatri M. Bhandari, Vijay Sonawane, and Tushar Phadtare

4.1 INTRODUCTION

The rapid evolution of wireless systems has transformed today's society and influenced how people interact and communicate with each other, as well as interacting with machines and intelligent devices, and the movement toward 6G (sixth generation) networks will create new types of communication that involve terahertz transmission, holographic connectivity, integrated sensing, and ultra-low latency networking. The need for this new level of capability presents a unique challenge to develop and implement an intelligence-based and more efficient spectrum allocation framework, using real-time, dynamic allocation of limited radio resources.

Traditionally, the command-and-control approach has been used for the allocation of spectrum, where regulatory authorities assign frequency bands to licensed operators on a long-term basis. While this method worked effectively for legacy systems, the end result is that many valuable and very useful frequency bands are not adequately or fully utilized and have significant waste in their fragmentation. Studies conducted by

DOI: 10.1201/9781003731597-4

the Federal Communications Commission (FCC) and others around the world have found that a majority of licensed spectrum remains unused (60–70%) [1] in many locations. As there is an increasing demand for wireless and mobile data (i.e., IoT devices and autonomous systems), this waste is leading to increasing pressure on existing systems to provide new applications and services, and thus may become the biggest bottleneck to the continued growth of existing systems.

The growing use of AI, ML, and blockchain technologies will continue to reshape the spectrum management landscape. The technologies of AI and ML provide Dynamic Spectrum Access (DSA) with predictive analysis and adaptive/automated decision-making, while blockchain can provide decentralized trust and transparent transaction histories between operators and users [2, 3]. Collectively, these technologies offer the building blocks for Intelligent Spectrum Management, which will be an important part of the coming 6th Generation (6G) World.

Motivation:
As stated previously, it is anticipated that the 6G networks of the future will deliver peak data rates above 1 terabit per second (Tbps), sub-millisecond latency, and massive connectivity to billions of devices [4]. To achieve these levels of performance, a dynamic and adaptable spectrum ecosystem must exist that is capable of self-organization, learning, and coordinating between many heterogeneous environments: Gen. Terrestrial, Aerial, and Satellite. The current static allocation methods will not be able to support this type of complexity; therefore, AI-driven dynamic spectrum management has been proposed and is expected to be the future of spectrum management and will use real-time data from cognitive radios, sensors, and network Analytics to make optimal allocation decisions [5].

Within this framework, blockchain technology provides accountability and fairness through the establishment of transparent contracts for the sharing of spectrum while also limiting malicious interference and misuse of spectrum [6]. Therefore, all of the mentioned characteristics of blockchain create a trustworthy, data-driven, and decentralized spectrum economy, which provides an efficient and governed model.

The objectives and scope of this chapter focus on the key conceptual and foundational elements of AI-based spectrum management for 6G wireless networks.

4.1.1 Objectives and Scope

The key objectives include:

- Identifying how spectrum management paradigms have evolved over the years and where current regulatory and technical frameworks are insufficient.
- Defining the architectural and algorithmic foundations of AI-enabled dynamic spectrum systems.
- Discussing how blockchain technology establishes the virtues of trust, transparency, and automation for decentralized spectrum marketplaces.
- Highlighting regulatory, ethical, and interoperability challenges that must be overcome prior to widespread adoption.

This chapter is structured to progressively develop the foundations and implications of AI-driven spectrum management for 6G networks. Section 2 traces the historical evolution of spectrum governance, beginning with command-and-control frameworks and progressing toward intelligent and decentralized paradigms. Section 3 critically examines current spectrum utilization challenges, including fragmentation, interference, regulatory rigidity, and economic barriers. Section 4 establishes the technical foundations of AI-based spectrum management by discussing machine learning models, deep learning (DL) techniques, federated intelligence, blockchain integration, and multi-agent cooperation mechanisms. The subsequent section addresses regulatory and ethical considerations associated with autonomous spectrum governance, emphasizing transparency, accountability, and policy harmonization. Finally, the chapter concludes with emerging research directions, including scalable explainable AI models, quantum-edge integration, sustainable spectrum frameworks, and cross-domain coordination. Together, these sections provide a comprehensive and forward-looking framework for intelligent, secure, and adaptive spectrum ecosystems in the 6G era.

4.1.2 Conceptual Overview of AI-Driven Spectrum Management

Due to the inherent shift in paradigm from traditional/rule-based spectrum management methods (of which there are many) to intelligent/data-centric/governance methods, AI-driven spectrum management is a significant change in how the radio spectrum is managed today. An AI-based spectrum management model utilizes a combination of real-time environmental data collection (sensing), context-based machine learning (learning based on experience) and decentralized decision-making (decisions made by multiple parties in a decentralized manner) to provide dynamic allocation of available radio resources in near real-time. While spectrum management methods of previous generations (e.g., 4G and earlier) relied on static pre-assigned frequencies or bands, the dynamic nature of AI-based systems allows them to continuously adjust to fluctuations in traffic demands, changing interference patterns and various regulatory policies, ultimately delivering superior spectral (i.e., bandwidth) efficiency and resilience compared to conventional spectrum management systems.

The conceptual framework, which is represented in Figure 4.1, is made up of four connected layers. The first layer is the data-sensing layer, which includes user equipment (UE), IoT devices, and base stations, all of which collect spectrum data from their local areas. The second layer is the "AI/ML Engine" layer. The AI/ML Engine utilizes advanced algorithms like deep reinforcement learning (DRL) and convolutional neural networks (CNNs) to predict how much of the available spectrum will be used at given times of day, locate interference, and optimize channel assignments by making predictive decisions based on data collected by layers 1 and 2. The third layer is the blockchain layer, which provides a trusted and tamper-proof method of documenting and tracking any access to the spectrum and the associated transactions. The use of smart contracts allows for the ability

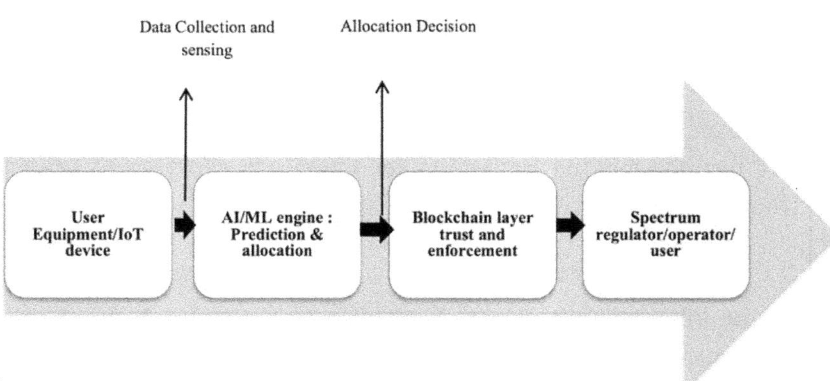

FIGURE 4.1 Conceptual model of AI-driven spectrum management.

to accept or deny a user's access to the spectrum based on the contents of the access agreement (or smart contract). The final and most top-level layer is the governance layer, where regulators and service providers are responsible for ensuring compliance with access rules, regulations, and policies [7].

With the above-mentioned multi-layered approach, the network can continuously learn from the generated historical data, make predictions about future conditions, and self-adjust in near real-time. The continuous feedback loop between the AI Engine's decisions and generated model will refine the predictive accuracy of the model and help to avoid conflicts among multiple users residing in the same area. The combined integration of blockchain technology will make all transactions occurring on the spectrum register independent, verifiable, and tamper-proof, thus establishing trust between multiple unrelated parties without a single point or entity of authority [8, 9].

The synergy between intelligence and transparency creates a shared cognitive asset, a spectrum that uses its "awareness" to adjust to the current situation and to conform to global objectives like: Equity, Efficient use of Energy, and Quality of Service (QoS). The shift to a new economic model creates opportunities for innovative ways to buy, sell, or lease radio frequency spectrum to other businesses and individuals using blockchain-based smart contracts, enabling businesses to create an obligation for payment in exchange for the spectrum. This means that in the case of an auction type system, the business would have full control over its spectrum, including the ability to keep the amount of the spectrum purchased or leased after an auction [10, 11].

As an example of the many benefits of this developing new paradigm, Table 4.1 offers a side by side comparison of the features of traditional spectrum management and spectrum management based on artificial intelligence (AI). The comparison shows the benefits in terms of adaptability, speed (latency), and efficiency of the spectrum's utilization. Blockchain technology will also ensure the auditability and trustworthiness of transactions. As a result of the combined effects of AI and blockchain, the spectrum ecosystem will be self-organizing, trustworthy, and highly efficient because all entities in the network contribute to the collective intelligence and capabilities of the system. This type of intelligent framework is likely to form the basis for the development of 6G communication systems. Supporting the vision of autonomous, adaptive and sustainable wireless environments.

TABLE 4.1 Comparison Between Traditional and AI-Driven Spectrum Management Models

Parameter	Traditional Model	AI-Driven Model
Allocation Type	Static, fixed frequency assignments	Dynamic, context-aware allocation
Decision Process	Manual	Automated
Adaptability	Low	High
Latency in Decision-Making	High	Low
Interference Mitigation	Reactive after detection	Predictive and proactive via AI models
Transparency and Trust	Limited visibility	Enhanced through blockchain ledgers
Utilization Efficiency	30~40 %	85~90%
Scalability	Constrained by central control	Highly scalable via distributed learning

4.1.3 The Paradigm Shift toward Intelligent Spectrum Governance

The continuous evolution of wireless technology constantly evolves the spectrum ecosystem and creates a more dynamic and complex environment than what existed in older generations of networks, which primarily used their spectral resources for voice and SMS services. In contrast, 6G has been designed to provide machine-to-machine communications; virtual reality experiences; tactile internet; and the use of AI to automating various processes or functions.

The rapidly evolving use cases of wireless technology result in complicated requirements concerning latency, reliability, and energy efficiency. This evolution in the wireless spectrum has removed any need for manually coordinating the wireless spectral resources. The cognitive functions of perceiving, learning, and reasoning about the status of a spectrum are provided by AI and ML technologies, which enable independent spectrum decision-making. The application of reinforcement learning, deep neural networks, and federated learning to DSA has resulted in significant improvements in DSA, the reduction of interference, and optimizing resources by AI. Machine learning algorithms implement optimal operational policies for ongoing interaction with the operational environment of the radio network, and a federated model maintains user privacy while allowing users to collaborate in training and share a functioning model without sharing their individual datasets. The intelligent algorithms derived through AI and machine learning will serve as the brain of future radio networks transition from a fixed asset approach to an adaptable self-organizing resource approach.

4.1.4 Integration of Blockchain for Trust and Transparency

One of the challenges in the DBA approach to DSA is that there is a need to build trusted relationships between the different participants in the DBA/RDO process such as the operators, regulators, and secondary users. DBA systems employ a traditional approach with a centralized database and as such can be subject to single-point failure, tampering or fail to synchronize between a numbers of database systems. The use of blockchain provides a secure, distributed database system that maintains an undistorted historical record of all of the transactions made for all spectrum access or leases and provides a basis for trust (in a DB) in a DBA system. In addition, blockchain networks that implement smart contracts provide the opportunity for users of spectrum to enforce the sharing rules associated with spectrum and to eliminate the need for intermediaries. In the case of a cognitive radio user requesting a temporary block of frequencies that are allowed to be used, the smart contract built using blockchain code can autonomously check whether that frequency is available, process the payment, and update the ledger on the decentralized sensing layer (the 'blockchain') in real time [11]. The decentralized approach not only increases security and traceability but also encourages spectrum sharing by providing economic incentives via tokenization, which encourages development of a new market model for space. Figure 4.2 depicts a conceptual blockchain-enabled model for spectrum co-ordination (immediately preceding this section), where multiple distributed nodes (across several protocols) collaborate to validate allocation transactions.

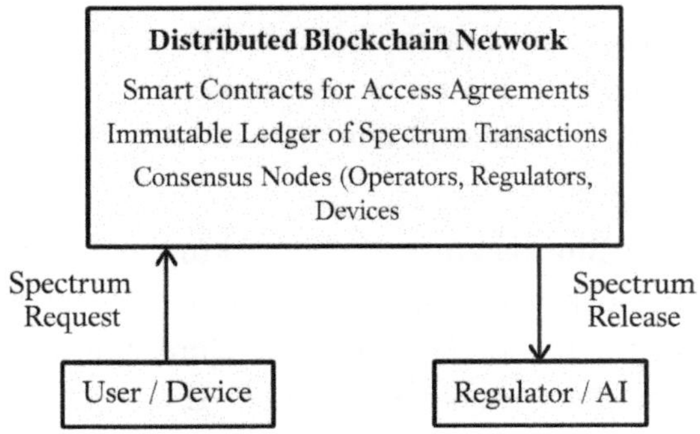

FIGURE 4.2 Blockchain-enabled spectrum sharing framework.

4.1.5 Challenges in the Transition to AI-Driven Spectrum Management

Although AI-based and blockchain-enabled approaches hold great promise for improving the efficiency and agility of spectrum management, they also present significant challenges in comparison to existing static management methods. The transition from traditional static spectrum allocation approaches (where frequency usage is predetermined) to intelligent and decentralized allocation approaches creates an overwhelming complexity of technological, regulatory, ethical, and operational issues that are not currently addressed by static allocation methods. As we implement the expected benefits of 6G technology, we must systematically address these multiple dimensions to provide performance and trust in future 6G systems.

While the integration of AI and blockchain technologies into spectrum management has great potential, it brings forth a variety of technical and regulatory challenges that must be resolved. The following are some examples:

- **Lack of Data Availability and Quality:** AI algorithms require vast quantities of labeled, quality spectrum data, which due to privacy and regulatory issues are not typically available.
- **High Computational Load:** The use of real-time AI processing means that high-performance computing equipment will be required at the "edge," thus leading to an increase in latency and/or power consumption.
- **Interoperability:** Ensuring interoperability between a diverse array of devices, networks, and jurisdictions remains one of the largest obstacles.
- **Explainability and Ethics:** Because AI can influence decision-making, decisions made by AI must have the ability to be interpreted and audited as compliance with established legal and ethical obligations.
- **Scalability of Blockchains:** When used in a public blockchain context, public blockchains may exhibit limitations on throughput and have higher transaction latencies. Therefore, alternative (optimized) consensus mechanisms are required [12-15].

Multidisciplinary collaboration involving researchers, policymakers, and practitioners in the industry will be necessary to address these challenges. It is critical to create and sustain an optimal synergy between AI and blockchain to achieve optimal performance, security, and sustainability within next generation spectrum ecosystems [16-18].

4.2 CHALLENGES IN CURRENT SPECTRUM UTILIZATION

The ability to effectively use the radio spectrum remains one of the most difficult tasks for modern communication written with regard to wireless technology's rapid advancement. As more wireless devices come into existence, new technologies and applications such as augmented reality and self-driving cars, as well as the number of types of access technology, all have put unprecedented demands on finite resource of spectrum[19-20]. Spectrum sharing, cognitive radio, and improvements in technology have reduced the scarcity of spectrum in part by providing greater access opportunities, but fragmentation, inefficient allocation schemes, lack of coordination across heterogeneous networks, and static policies also create significant inefficiencies with respect to the uses of radio spectrum [21-22]. The inefficiencies of the current methods of managing the radio spectrum impede any possibility of bringing new technologies and applications to market in a timely manner. This section describes the major impediments to optimal use of radio spectrum today, and emphasizes why a shift to AI-based dynamic management represents a critical opportunity for the development of the next evolution of cellular technology 6G [23].

4.2.1 Spectrum Underutilization and Fragmentation

The greatest barrier to management of the spectrum is the unbalanced use of different frequency bands. While it appears as if there is a continuing growth of the overall amount of spectrum in general by the number of devices in the market, there are significant numbers of bands that are licensed to use that spectrum but are not actually being used or utilized over an extended period. This represents a lack of efficiency with respect to the spatial and temporal use of the radio spectrum. For example, in urban areas, where there is a large amount of cellular activity, the spectrum is being congested as a result. However, in rural areas or remote areas, there typically are very low occupancy rates in licensed frequency bands [24]. This creates a situation where, due to a lack of flexibility in regard to how spectrum is allocated and subsequently reassigned, there are

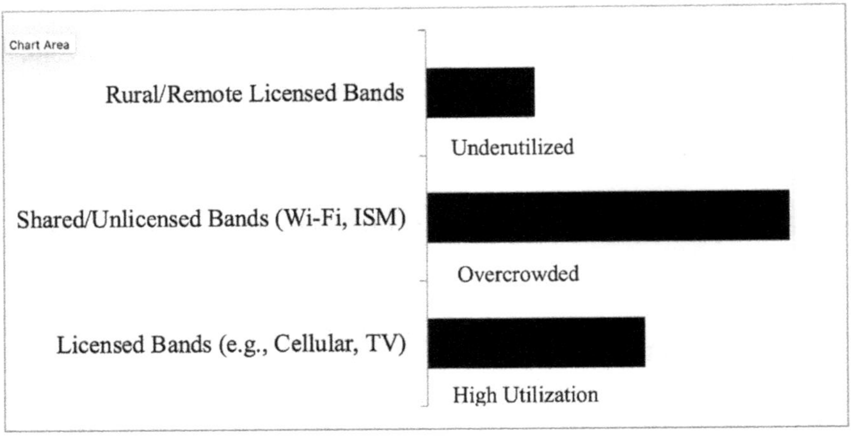

FIGURE 4.3 Illustration of spectrum fragmentation and underutilization.

substantial differences in the availability of spectrum between urban and rural locations.

The use of the radio frequency spectrum is extensive and fragmented, spanning many different sectors including: Cellular Communications, Broadcasting, Radar, and Satellite. Because of this fragmentation, it has been challenging for providers to manage interference and reuse spectra efficiently. Figure 4.3 shows that certain bands are significantly oversubscribed while others are underutilized. Today's approach to allocating allocations of the radio frequency spectrum by services or service providers is ineffective since the majority of allocations are static and, therefore, unable to accommodate real-time fluctuations in demand for the spectrum leading to an estimated global spectrum inefficiency factor of 20% to 40% [25].

4.2.2 Interference and Coexistence Issues

As wireless devices have continued to multiply at an exponential rate over the past several years, managing interference from these devices has become increasingly complicated due to the fact that these devices are likely to use the same frequency ranges (e.g., the 900 MHz, 2.4 GHz and 5 GHz bands) which creates opportunities for Interference among wireless devices and degrades the perceived QoS of each device [26]. Currently, the industry relies heavily on methods such as fixed frequency planning, guard bands, and centralized coordination for interference mitigation; however, these methods will not be sufficient for 5G/6G heterogeneous environments (e.g., smart homes, Internet of Things (IoT), satellite, and connected vehicles) since the technology and devices will be very diverse. As a result, the increasing congestion of unlicensed frequency bands (e.g., the 2.4 GHz and 5 GHz bands) will cause multiple collisions, resulting in lowered data rates and reduced QoS. In addition to these issues, the lack of adaptive or predictive interference control solutions will pose a serious challenge for maintaining higher levels of QoS in dense deployments.

4.2.3 Inefficient Spectrum Sharing Policies

Static spectrum sharing strategies are a continuation of fixed policies that cannot change. The licensed shared access (LSA) and citizens broadband radio service (CBRS) models, for example, require central databases to enable the management of priority levels and the authorization of access. While centralized databases are a better than the use of the fixed allocations in terms of sharing Efficiently, they still rely on a manual process for updating the operational rules and also the situational awareness is limited. Therefore, the actual sharing decisions made with these systems do not take into consideration real-time conditions, including mobile devices, and variations in traffic conditions. Centralized control creates further issues by creating a single point of failure and hindering the ability to scale. Without AI-based mechanisms capable of learning and automatically adjusting their operation, spectrum sharing systems will continue to exhibit inefficiencies in their operation.

4.2.4 Regulatory and Administrative Constraints

The regulations for spectrum cannot be standardized and are split between different jurisdictions. Every country has a different licensing standard, compliance standards, and different enforcement mechanisms, which limits cross-border coordination, especially for the satellite and aerial networks that have been identified within Global 6G systems [27]. The process for license proposal approval for new frequency allocations is often lengthy and fraught with bureaucracy which limits the development and testing of new innovative access methods. For example, the spectrum reallocation for 5G mid-band frequencies required significant negotiations between the defense, broadcasting, and telecommunication agencies for several years. To ensure true global interoperability, spectrum regulatory frameworks need to evolve to include AI-based monitoring, dynamic compliance monitoring, and the use of blockchain for audit purposes to increase transparency and speed of the administrative process.

4.2.5 Lack of Real-Time Spectrum Awareness

Correct situational awareness, the ability to continuously monitor the spectrum over a geographical area and in multiple frequencies and times, is a fundamental requirement for efficient use of spectrum. Presently, existing sensing systems are inadequate in providing full coverage of the entire spectrum in terms of both time of measurement and space of coverage. Most regulatory agencies continue to use periodic measurements with a fixed monitoring station, thus missing the fast-paced variability in potential occupancy. As a result, the decisions made using these old or incomplete data may adversely affect how that spectrum will be managed. New methods of sensing such as crowd sourcing, cognitive sensor networks, or software-defined radio, while they may provide some degree of improvement in the measuring or monitoring of spectrum usage, do not coordinate across the globe. If these new spectral sensing capabilities are combined with AI-driven predictive analytics, it will be possible to proactively manage how that spectrum is utilized by forecasting the occurrence of spectrum 'holes' prior to them actually occurring [28].

4.2.6 Economic and Operational Barriers

In addition to the technological barrier to efficient spectrum management, there are also economic and operational issues that contribute to the problem. The high cost of acquiring a spectrum license acts as a significant barrier for many smaller organizations (particularly startups and rural operators) from entering the marketplace or to operating effectively within it. Thus, significant frequency resources remain underutilized in areas that are likely to be less profitable for the operators. Additionally, impeding spectrum sharing among operators is the lack of trust between operators and the inability to develop or consistently apply a revenue sharing model. By removing these barriers through the use of a blockchain issued smart contract, it would be possible to eliminate the adversarial relationship between the spectrum owners and those looking to lease that spectrum, therefore dramatically improving the efficiency of spectrum trading. This technology and application has yet to be fully developed and will remain within the research domain until it can accommodate all elements of current spectrum trading processes and have sufficient business incentives to ensure large-scale adoption.

4.2.7 Hardware and Sensing Limitations

Beyond policy, coordination, and algorithmic challenges, practical RF hardware constraints significantly influence the effectiveness of spectrum management. Spectrum sensing accuracy is limited by the physical characteristics of sensing devices. Software-defined radios exhibit constrained dynamic range (typically 60–80 dB), phase noise that affects interference detection, and sampling-rate limitations governed by the Nyquist criterion, which restrict wideband monitoring. Front-end saturation caused by strong interferers further degrades sensing reliability. In addition, spectrum sensors require periodic calibration, as temperature variations and component aging alter RF sensitivity over time. Energy consumption also presents a major limitation, since wideband sensing can require hundreds of milliwatts of continuous power, making persistent monitoring infeasible for battery-powered IoT nodes. Consequently, AI-based spectrum management systems must explicitly account for imperfect measurements, energy budgets, and coverage gaps. Federated and cooperative learning approaches mitigate these constraints by aggregating observations from multiple heterogeneous sensors rather than relying on idealized centralized measurements. Table 4.2 summarizes the major technical, regulatory, and operational challenges associated with current spectrum utilization, together with their corresponding implications for next-generation AI-driven wireless networks.

TABLE 4.2 Major Challenges in Current Spectrum Utilization

Challenge	Description	Implications
Underutilization and Fragmentation	Uneven occupancy across regions and services	Poor efficiency and wasted resources
Interference and Coexistence	Overlapping transmissions in dense areas	QoS degradation and connectivity drops
Static Sharing Policies	Fixed rules and centralized control	Inflexibility and latency in access decisions
Regulatory Limitations	Slow coordination and policy mismatch	Delayed innovation and interoperability gaps
Limited Spectrum Awareness	Lack of real-time data on occupancy	Inefficient allocation and poor prediction
Economic Barriers	High licensing costs and weak incentives	Restricted participation and innovation lag

Spectrum management currently operates on a reactive, rigid, and fragmented basis with limited flexibility to fulfill the performance, latency, and reliability needs of future 6G technology. The only way to address these issues is through employing new management systems that utilize AI, ML, and blockchain technology. This type of management system transforms the way that spectrum is managed by using data to drive decisions and allocate spectrum resources in an optimal, secure, and equitable manner across the user base, devices, and geographies.

4.3 FOUNDATIONS OF AI-DRIVEN SPECTRUM MANAGEMENT

Historically, the evolution of communication technologies from their inception to their present status has experienced many transformations, but it seems that those changes were pretty much all result of the method through which we transform our communications. Now, however, it appears the most significant transformation in communications will be that between traditional management systems to those of intelligence-based spectrum management, AI-SM for short, where the introduction of AI is likely to continue on in its use over the coming years. Due in part to the ongoing growth of 5G technologies and their eventual transition into 6G Systems, AI-SM will become essential in addressing the need for rapid adaptability to real-time changes, maximizing resources, and building the ability for continuous learning. AI-SM no longer uses established rules and policies but instead implements algorithms and tools to provide the best use of available data. AI-SM will continually refine and update their decisions based on constantly dynamic environmental conditions and will present the most appropriate spectral radiative frequency bands for continued transmission. The section in this chapter will address the major elements of the AI-SM paradigm, the key structural components and processes, and how to use ML, DL, and blockchain technologies for operating securely and transparently.

4.3.1 Conceptual Framework of AI-Driven Spectrum Management

Artificial intelligence spectrum management (AI-SM) uses cognitive intelligence to model a human's thought process when making decisions regarding wireless networks. AI-SM collects information via sensors, base stations, and device endpoints to continually adjust the policies that are defined at a specific moment, and is responsive to both time and geography changes in the use of the wireless spectrum [29].

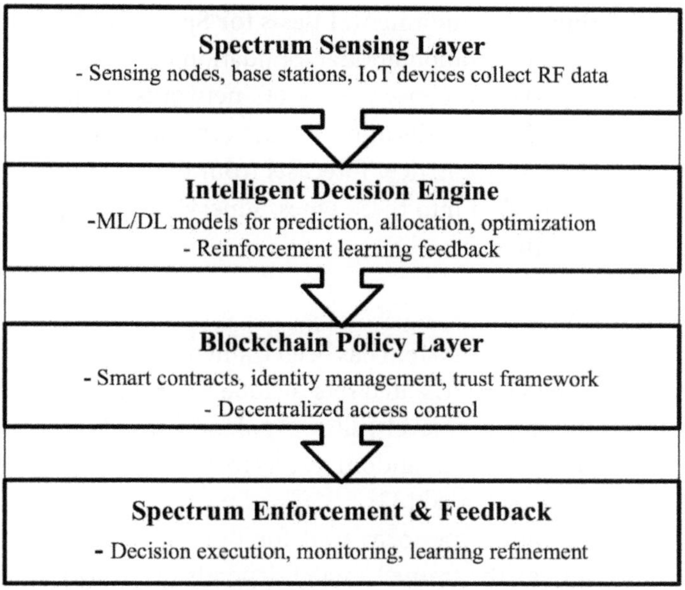

FIGURE 4.4 Core architecture of AI-driven spectrum management.

The core architecture of AI-SM consists of four interconnected components (illustrated in Figure 4.4):

- Spectrum Sensing and Data Collection: Continuously monitoring spectrum use, interference, and signal power levels and geolocation information.
- Intelligent Decision-Making Engine: Makes predictions concerning future spectrum requirements through advanced ML and AI algorithms to optimize allocation practices.
- Blockchain-Aided Policy Layer: Provides a level of trust, transparency, and traceability among participating parties through distributed ledgers (DLT) and smart contracts.
- Reinforcement and Federated Learning Mechanism: Improves continually through using past decisions as the basis for making future decisions.

By combining these four modules together within a single architecture, AI-SM enables the development of intelligent 6G networks by allowing the system to be contextually aware, organized, and work autonomously as required [30].

Machine Learning as a Fundamental Basis for Spectrum Management

Machine learning is the computational foundation on which AI spectrum management (AI-SM) and consequently 6G networks are built. Machine learning facilitates the identification of patterns within large datasets, enabling systems to derive logical forecasts from these patterns. Learning can generally be classified into three types: supervised, unsupervised, and reinforcement learning [31].

Supervised Learning

In supervised learning, the system has been trained to recognize patterns by using pre-labeled datasets. The datasets include several patterns of signals, interference, and occupancy levels, allowing algorithms like support vector machines, decision trees, and random forests to be used to classify channels as either available or occupied [32].

Supervised techniques perform well for both spectrum sensing and anomaly detection, as the trained model depends on historical transmission data to detect any anomalous (unauthorized) transmissions. However, when the radio frequency environment has changed substantially, additional supervision will have to occur on an ongoing basis, meaning retraining becomes a very expensive computational effort.

Unsupervised Learning

Unsupervised machine learning techniques utilize a different type of training where the training examples provided to the machine learning model are not labeled. With unsupervised techniques, a machine learning model is able to extract and identify patterns from the radio frequency data [33], allowing for the use of unsupervised machine learning techniques, such as classification and clustering. The technique known as clustering allows for the clustering of frequency bands that exhibit similar behavioral characteristics, thus facilitating the reuse of channels in a more efficient manner. While unsupervised learning approaches provide flexibility, it may be necessary to conduct additional post-processing to improve the accuracy of decisions made.

Reinforce Learning

Reinforce learning, the specific approaches of Q-Learning and Deep Q-Networks, represents the most advanced approach to dynamically allocating the radio frequency spectrum. In reinforce learning; an agent interacts with its environment through a series of actions (i.e., assigning

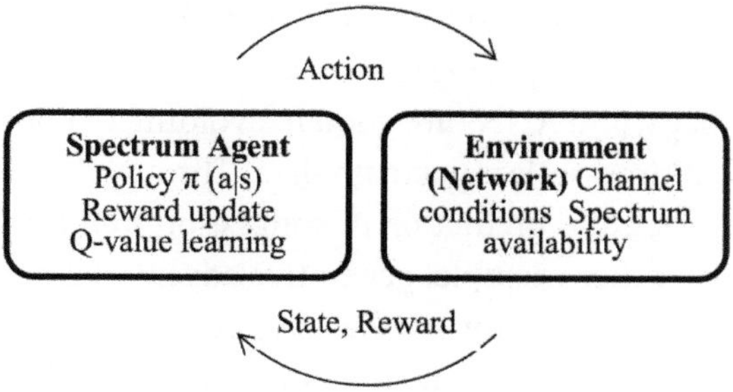

FIGURE 4.5 Reinforcement learning loop for dynamic spectrum allocation.

or releasing a channel) that ultimately maximize the Agent's reward function, typically measured by throughput, fairness, and energy efficiency [34]. Reinforcement learning enables self-learning behavior without prior knowledge, making it ideal for non-stationary environments. An example would be a cognitive radio using reinforcement learning techniques which would learn how to avoid interference and also maintain a high level of spectral efficiency all on its own. The agent-environment interaction loop is shown in Figure 4.5.

4.3.2 Federated Learning for Collaborative Spectrum Intelligence

Federated learning is one of the most recent methods for managing spectral resources. Federated learning allows many users/devices in a 6G network to train a common global model together without sharing their local raw data, thus maintaining data privacy and adjusting the amount of communication overhead, since only updates to the model (not the raw data) will be shared [35]. In a large-scale 6G network, base stations will be able to develop their models using their own local spectrum data, and each base station will be contributing to the same global model maintained by either an operator or regulatory agency. Federated learning methods provide the additional benefit of improving security since there is no single location to store all of the data.

4.3.3 Role of Deep Learning (DL) in Spectrum Analytics

DL provides an extension on traditional ML techniques. DL uses multi-layer neural networks that are capable of automatically abstracting high-level features from low-level data, such as raw signal sources. A number of different techniques within DL, including CNNs and RNNs (recurrent neural networks), have produced good results in the areas of spectrum sensing, modulation recognition, and interference classification [36]. CNNs are especially good at recognizing spatial patterns within the power spectral density (PSD) maps, and RNNs are good in avoiding temporal dependency between different time-series spectrum data. Through the identification of spectrum occupancy and prediction of congestion trends can theoretically be achieved. As another example, generative adversarial networks (GANs) have become a popular way for synthesizing realistic spectrum data. With GANs, modeling can be done on situations when it is difficult or impossible to obtain enough data points for labeling purposes [37].

4.3.4 Blockchain Integration for Secure Spectrum Governance

To establish a secure and equitable environment for transactions using spectrum, AI-based systems need a foundation of trust. This trust is provided by Blockchain technology: the use of immutable ledgers and smart contracts to record all allocations, returns, or payments made in relation to access to the spectrum.

The blockchain-based framework defines:

- Policy compliance and licensing agreements via smart contracts enabling automation;
- Elimination of centralized authorities via decentralized consensus;
- Enabling accountability and deterring fraudulent use via identity management.

For example, when an operator wishes to gain access to a temporary frequency band, they can request access via a smart contract which will automatically verify compliance, allocate the frequency channel to them and record the transaction for them; all without any human involvement. This transparency also creates a stronger sense of security and trust between operators, as well as an ability to audit, in real-time.

4.3.5 Multi-Agent and Cooperative Learning Models

Future sixth-generation (6G) networks will consist of billions of connected devices. Managing this level of complexity requires a multi-agent learning approach, where multiple AI agents can work in cooperation or competition for shared spectrum resources, with each agent representing one of the elements in the network (base station, vehicle, or satellite) while being able to collaborate with the others. The agents will work together to maximize local objectives such as throughput, delay and power consumption, while still meeting the requirements of the collective, and therefore maximizing the performance of the entire combined network [38]. Cooperative multiagent systems can realize joint optimization through the use of game-theoretic and reinforcement-based (i.e., learning) methods where agents share information with each other in real-time and use that information to adapt their behaviour in response to their interactions with the other agents. The decentralized nature of the coordination of cooperative multiagent systems is similar to how biological ecosystems operate; hence, they will be stable and will adapt to changes in the operating environment.

Spectrum management using AI is an integrated approach to creating intelligent wireless ecosystems. Through this method, three technologies are brought together (machine learning, DL, and distributed ledger technology) to create a foundation of intelligent wireless network ecosystems.

AI Spectrum Management will allow for:

- Adaptability to changing conditions, real-time decision-making, and flexibility in programming/strategy through reinforcement learning.
- Preservation of user data confidentiality through federated and cooperative data models.
- Creation of trust and transparency through blockchain technology and using the blockchain as an auditable record.
- Continuous refinement of AI decision-making strategies through self-evolving intelligence.

Together, these capabilities represent the first step towards 6G networks that are autonomous, self-regulating and optimize spectral utilization across multi-domain infrastructures in a fair and secure manner.

4.3.6 Practical Deployment Considerations

Although AI-based spectrum management frameworks demonstrate strong theoretical performance, practical deployment introduces additional constraints. Model complexity must be balanced against decision latency, particularly in ultra-reliable low-latency communication scenarios where sub-millisecond response is required. Deep architectures offer higher prediction accuracy but may introduce processing delays, making lightweight models more suitable for edge-based allocation tasks. Data availability is another constraint, as supervised approaches depend on sufficiently labeled

TABLE 4.3 Key Performance Metrics for AI-Driven Spectrum Management

Metric	Definition	Significance
Spectral Efficiency (η)	Ratio of utilized bandwidth to total available spectrum	Measures overall utilization effectiveness
Decision Latency (τ)	Time taken for spectrum sensing and allocation	Indicates real-time responsiveness
Learning Accuracy (α)	Percentage of correct decisions by the AI model	Evaluates reliability of prediction
Energy Efficiency (E)	Energy consumption per allocation cycle	Assesses sustainability and scalability
Blockchain Latency (Lb)	Delay introduced by distributed consensus	Determines transaction speed and scalability

spectrum datasets; transfer learning and simulation-assisted training can reduce this requirement. Hardware limitations, including restricted power budgets of edge devices and computational capacity constraints, necessitate model compression techniques such as quantization and pruning. Furthermore, spectrum environments are inherently non-stationary, requiring periodic retraining and drift detection mechanisms to maintain allocation reliability. Consequently, deployment-ready AI systems must integrate latency awareness, adaptive learning, and hardware-conscious optimization to ensure consistent and scalable performance in operational 6G networks. Table 4.3 presents the key performance metrics commonly used to evaluate AI-driven spectrum management systems, highlighting their definitions and significance in assessing network efficiency, responsiveness, learning performance, energy consumption, and blockchain scalability.

4.4 KEY ENABLING TECHNOLOGIES (AI, ML, AND BLOCKCHAIN INTEGRATION)

Next-generation spectrum management represents a convergence of three technologies (AI, ML, and blockchain) that are all creating a new and intelligent data-driven, secure foundation to allocate and use radio frequency spectrum dynamically. Traditional spectrum management methods are based on rigid policy and centralized control. By integrating AI, ML and blockchain, wireless networks can sense, learn, adapt, and govern themselves autonomously. This section discusses each of the pillars of AI spectrum management and how together they create a self-organizing and trustworthy ecosystem for AI-driven spectrum management.

4.4.1 AI as the Cognitive Core

Spectrum management intelligence is the cognitive component of future intelligent spectrum management and provides a mechanism for creating awareness, reason, and decision-making capabilities which allow the network to act on its own and have more flexibility than just following a strictly defined set of rules. With access to vast amounts of real-time data from sensor networks (sensing nodes, base stations), and IoT devices, intelligent spectrum management systems utilize AI technologies to recognize and evaluate the condition of its environment, including interference, demand, and signal strength variation [39]. Intelligent spectrum management uses adaptive logic and probabilistic reasoning techniques to allow for context-based decision-making, with examples including the ability to dynamically change between frequency bands, adjust transmission power, and choose optimal frequencies for minimal interference. Intelligent spectrum management systems' capability to both sensorize their Environment and react accordingly elevates the reliability of operating rate, the efficiency in operation, and the environmental sustainability of the system. Moreover, AI has predictive capabilities, enabling AI-based intelligent spectrum management systems to be more reliable by identifying traffic congestion patterns and peaks before they occur. Through the pre-allocation of resource to address congestion, intelligent spectrum management systems can create an environment where users do not have to reactively manage the spectrum but rather proactively manage it, increasing spectral efficiency in dense urban environments.

To summarize, AI enables autonomous sensing, reasoning, and optimization capability in the modern date era of wireless technology, forming a full, autonomous network of communications. These systems build on existing machine learning algorithm and blockchain technologies and integrate them to provide complete, intelligent communication network infrastructures [40].

4.4.2 Machine Learning as the Data-Driven Decision Engine

Machine learning acts as an enabler for AI by translating raw spectral data into usable insights and adaptive strategies. Machine learning algorithms are able to categorize, classify channel states, and provide predictions of channel availability based on changes in network conditions. Unlike traditional systems that create rules based on fixed inputs and create fixed outputs, machine learning systems will continue to learn and improve their ability to make decisions and generate insights over time [41]. Using supervised and unsupervised learning techniques, spectrum management systems are capable of determining which channels are idle, detecting anomalies and clustering channels into frequency bands based on the type of user activity that has occurred over time. The data-driven nature of these techniques enables the adaptive development of access policies as real-world circumstances change, giving optimal access to available bandwidth. As demonstrated by recent research, hybrid machine learning techniques can reduce false detection rates and increase the accuracy of identifying channels by more than 25% when compared to the conventional approaches to sensing [42]. Within the area of machine learning, the subset known as reinforcement learning provides a method for agents to develop and adapt behaviors through continuous interaction with their environments. The agents are rewarded for their efforts to minimize any interference from others while maximizing the amount of data they send through the spectrum. By using this system, agents can develop their own optimized policies on how best to use the spectrum without needing an explicit supervisor [43]. The use of reinforcement learning helps establish the foundation for autonomous decision-making, thus enabling spectral management systems to be self-managing and self-regulating. Federated learning will allow multiple independent edge devices to cooperate with each other during training of the machine learning models. Each edge device can maintain its own data privacy while decreasing communications required between edge devices [44]. Distributed learning strategies are key to creating a secure and efficient approach for providing decentralized intelligence in 6G networks across multiple nodes.

Therefore, machine learning will be the backbone of decision-making within the architectural framework of the overall system and enable the development of intelligent means of working, the development of flexible frameworks for future and continual enhancement with past and present data.

4.4.3 Blockchain as the Trust and Transparency Layer

The implementation of blockchain in decentralizing the management of frequency bands adds a layer of security and establishes the levels of trust required. Since managing and using the frequency spectrum involves multiple parties: operators, regulators, enterprises, and individuals, blockchain provides a way for all parties to coordinate transparently and report verified transactions to each participant. The record or ledger in blockchain will maintain a record of every activity associated with the frequency spectrum – allocation, release, or negotiating contracts for trading – and this record will be maintained in an immutable database, which provides complete traceability of every operational action. Therefore, the blockchain ledger prevents any malicious activities through the housing of transactions in a decentralized location where unauthorized access and manipulation will be easily identified while also providing a means for accountability and adherence to FCC regulations [45]. The smart contracts available within the blockchain enable the automation of implementing policy and licenses. For example, if a network operator wants to use a frequency band, through a smart contract, the blockchain will validate the network operator's eligibility based on established qualifications and federal regulatory rules. When the smart contract has completed the operation, it will log the transaction and facilitate the release of the frequency band without any involvement from an administrator or the possibility of human error [46]. Decentralized frequency trading can also leverage blockchain to enable participants to create a token for-and-sell frequencies like any other digital asset via secure platforms [47]. The marketplace provides for quick and efficient redistribution of frequencies that might otherwise go underutilized by competitors, thus creating the maximum efficiency possible. The combination of blockchain, AI, and ML assures that each intelligent decision made by an autonomous agent or predictive model will be auditable, tamper-proof, and will have a verified layer of trust due to the cryptographic integrity and decentralized consensus of the blockchain, thereby greatly reducing any potential risk for bias, misuse, manipulation, or malicious interference from an outside source [48].

The framework is made up of three layers to provide the foundational ethical and regulatory structure for blockchain technology so that all participants in the AI-managed spectrum ecosystem will have equal access to securely and transparently share and use the spectrum via blockchain technology.

4.5 REGULATORY AND ETHICAL CONSIDERATIONS IN AI-DRIVEN SPECTRUM MANAGEMENT

The convergence of AI, ML, and blockchain technology in the management of spectrum has changed how spectrum can be managed. For decades, the governance of spectrum has been based on centralized regulatory bodies such as the International Telecommunications Union (ITU), the FCC, and the Telecom Regulatory Authority of India (TRAI), whose spectrum policies were based on the static allocation of spectrum for the long term under a license arrangement. As AI-based systems and other intelligent, decentralized systems continue to develop, regulation of the management of spectrum must continue to evolve to address the issues surrounding autonomy, accountability, and ethical transparency in algorithm-driven decisions [49].

4.5.1 Evolving Regulatory Paradigms

Historically, the regulation of spectrum has included a long-term exclusive license that allows the operator to have predictable methods of operation as well as protection from interference from other operators. With AI-based systems making the real-time dynamic adjustments to how spectrum is used, it is common for operators to use the same frequency in a different geographic or administrative area than originally granted by the regulator. Thus, regulators will need to implement flexible licensing and DSA frameworks that will allow for the contextual decision-making capabilities of machines rather than human intervention [50].

Regulators will also have to take into account new questions being raised about accountability, compliance, and auditing of AI systems:

- Who should hold the responsibility when an AI system allocates a frequency incorrectly?
- How will automated decisions made by AI systems be compliant with the provisions of both national and international rules regarding the protection of operations from interference?
- What types of auditing mechanisms must be in place to ensure that AI learning algorithms remain unbiased and that decisions made by AI remain transparent?

New initiatives from organizations like the ITU-R Study Group 1 on Spectrum Management and the IEEE P7000 Standard Series seeking to develop ethical AI have established global standards for how we view accountability and responsibility regarding the growing reliance on automated systems [51]. These organizations have established three core principles for developing trust in algorithmically managed spectrum resources; these principles include human oversight, explainability, and traceability. Furthermore, the rapid emergence of decentralized blockchain-based spectrum trading creates a new set of regulatory policy challenges (i.e., taxation, competition, jurisdiction of data), necessitating that regulatory bodies establish additional rules that ensure that all transactions based on decentralization technologies adhere to present-day standards while also fostering both innovation and fairness in market operations.

4.5.2 Toward a Responsible Spectrum Ecosystem

The key to achieving a responsible spectrum ecosystem lies in responsible technology innovation. This involves creating a technology that integrates technical advancement with both human values and societal needs. The three principles of a responsible AI-driven spectrum ecosystem are as follows:

- Transparency/Explanatory: All automated decisions must be explainable and verifiable by the regulatory authority.
- Accountability/Oversight: Developers and operators of intelligent systems should have clearly defined legal liability for their actions.
- Inclusiveness/Sustainability: Spectrum access policies must enable equitable user participation and seek to minimize the adverse effects on the environment.

Success in building and deploying intelligent spectrum networks will depend on international collaboration. To help enable this, it is important that organizations like International Telecommunication Union, 3rd Generation Partnership Project, and other relevant regional authorities work together to create interoperable policy frameworks that enable intelligent spectrum networks to operate legally and ethically between nations. As we move toward an AI-centric method for managing spectrums, we will create a new system where regulations will not disappear, but be reimaged. Therefore, human regulators will change from a focus on rule enforcement to a focus on ethical oversight to ensure that autonomous systems use AI effectively in the public interest, while at the same time maximizing efficiency and innovation.

For an AI-centric method for managing spectrums to be successful, we need to improve the technology, while at the same time, re-define the regulatory and ethical frameworks to best suit our needs as we move forward. The overall balance between automated action and human accountability must be clearly articulated to achieve a transparent, fair, and sustainable spectrum ecosystem. By embedding responsible design and international partnerships into our innovations at all levels, we will be able to create a cooperative intelligent spectrum ecosystem that has responsibility and integrity at its foundation.

4.6 FUTURE RESEARCH DIRECTIONS

4.6.1 Scalable and Interpretable AI Models

The remarkable capabilities of DL systems in the area of pattern identification and decision modeling have not yet been rivalled, due largely to their level of sophistication and uncertainty with regard to regulation acceptance. However, future work should be undertaken to investigate ways to create "Lightweight" and "Explainable" AI systems at the edge of a network, which will enable ease of use and provide transparency for audit proof. Federated transfer learning and knowledge distillation are two examples of methods that have been employed effectively to deliver scalable intelligence, without decreasing interpretability or increasing energy costs.

4.6.2 Integration of Quantum and Edge Intelligence

As quantum computing continues to grow in capability, combined with the introduction of Edge AI, the ability to enhance computational speed and accuracy of decision-making in real-time spectrum management may be greatly improved. For example, algorithms that employ quantum-assisted learning will be able to provide greater effectiveness and efficiency when optimizing multi-agent systems, while edge nodes can provide more localized intelligence and reduce latency and backhaul load. In this context, it is possible to develop new hybrid quantum-edge-AI concepts as the foundation for ultra-responsive 6G and beyond 6G networks.

4.6.3 Sustainable and Green Spectrum Governance

AI technologies and blockchain networks, while both very powerful, require significant amounts of energy to perform well; therefore, the next part of the research agenda should be the development of green spectrum management models which have minimal energy need for the training and running of AI models, and also the consensus process of blockchain. Use of energy-aware reinforcement learning and proof-of-stake methods can be implemented in these models, thereby reducing their carbon emissions and possibly lowering intelligent networks' overall carbon footprint substantially. Integrating renewable energy sources into spectrum infrastructure could further advance sustainability goals aligned with global ITU-T L.1470 recommendations on environmental efficiency.

4.6.4 Cross-Domain and Satellite Spectrum Coordination

The 6G network will evolve from current terrestrial networks to have a multi-domain (satellite, aerial/open air, and maritime) environment in which multiple layers of heterogeneous network operators will be operating with a coordinated spectrum plan. Because of the multiple heterogeneous layers of network operators and operators' equipment, there will be many difficulties in managing interference, managing synchronization, data sharing, etc. Cross-domain spectrum intelligence will provide the foundation for global synergy in the use of frequency across the Earth-space environment by leveraging both AI and advanced wireless technologies.

4.6.5 Ethical AI and Policy Harmonization

The first step to building an ethical and internationally compliant AI is to define AI's interoperable relationship with existing global legal jurisdictions. To develop an interoperable, ethical AI, basic research should be guided by a set of technical and ethical decision-making frameworks (e.g., IEEE, ITU, and OECD) that define the ethical/legal responsibilities of automated systems. Building an ethical AI involves developing standards for trust between providers and consumers of AI technology; audit/training history; and cross-border governance policies for data shared between both consumers and providers of AI technology.

CONCLUSION

There has been a paradigm shift in wireless communications with the evolution of AI-based spectrum management, moving from static governance rules to dynamic autonomous systems. This chapter explains how AI, ML, and blockchain work together to enable autonomous capable networks to sense, understand, and act on things in the world around them, through the development of 6G networks, with intelligence available everywhere. In addition to outlining the Foundations of Intelligent Spectrum Systems, the chapter also discusses the challenges associated with implementing AI, ML, and blockchain as technologies that will empower the future development of autonomous networks to function with greater vigilance and efficiency. Through the integration of explainable AI, quantum edge intelligence, and sustainable structures, AI will guide and direct the research agenda of the next decade. To successfully exploit and use these emerging technologies, we must strike a balance between using the 21st century technologies that we are developing, meeting our ethical obligations to society, and protecting the environment to create a wireless ecosystem that will flourish in the future.

Consolidating the above assertions will lead to a conclusion that AI will enable a fundamental shift from an architectural view to one where the network it-self thinks, learns, and governed itself in alignment with human values and global sustainability strategies. This will provide a foundation for creating autonomous, equitable, and intelligent communication systems, which will be the basis for 6G and beyond.

For researchers and practitioners implementing AI-driven spectrum management, several practical insights emerge from this chapter. Interpretable and lightweight learning models should be prioritized during early deployment stages to satisfy regulatory expectations and enable transparent decision-making. Transfer learning and simulation-assisted training can significantly reduce real-world data requirements, accelerating deployment in data-scarce environments. Latency–accuracy trade-offs must be carefully managed, particularly for ultra-reliable low-latency applications where edge-based inference is essential. Privacy-preserving designs, such as federated learning, are recommended to avoid centralized collection of raw spectrum measurements. When integrating blockchain, permissioned or consortium-based ledgers may be preferable to public chains due to lower transaction latency and energy consumption. Finally, early engagement with regulatory authorities is critical to ensure alignment with evolving spectrum policies and to facilitate smoother standardization and operational approval processes.

PRACTITIONER'S TAKEAWAY

If you are working in wireless network design or spectrum policy, the message here is hard to ignore. The old way of assigning fixed frequency bands is leaving 60–70% of licensed spectrum sitting idle while demand keeps climbing. AI and machine learning give networks the ability to sense, predict, and allocate in real time – without waiting for a human to intervene. Blockchain quietly does the unglamorous work of making those transactions auditable and trustworthy across competing stakeholders. The combination is not just a technical upgrade; it is a governance shift. Start thinking now about how your organization handles spectrum decisions, because 6G will not wait.

REFERENCES

[1] M. Nekovee, "Cognitive radio access to TV white spaces: Spectrum opportunities, commercial applications and remaining technology challenges," *Proc. IEEE DySPAN*, pp. 1–10, 2010.

[2] M. N. Islam, A. A. Rahman, and A. Anpalagan, "A machine learning approach to spectrum management in 5G and beyond," *IEEE Wireless Commun.*, vol. 27, no. 4, pp. 35–41, Aug. 2020.

[3] A. D. Wyglinski, M. Nekovee, and T. Hou, *Cognitive Radio Communications and Networks: Principles and Practice*, Academic Press, 2010.

[4] W. Saad, M. Bennis, and M. Chen, "A vision of 6G wireless systems: Applications, trends, technologies, and open research problems," *IEEE Netw.*, vol. 34, no. 3, pp. 134–142, May 2020.

[5] S. Choudhury et al., "AI-enabled dynamic spectrum sharing and coexistence for 6G," *IEEE Access*, vol. 10, pp. 110234–110250, 2022.

[6] Y. Zhang, H. Liu, and J. Zhang, "Blockchain-based decentralized spectrum management for 6G networks," *IEEE Trans. Netw. Serv. Manage.*, vol. 19, no. 2, pp. 1064–1076, 2022.

[7] N. Al-Falahy and O. Y. Alani, "Technologies for 5G networks: Challenges and opportunities," *IEEE IT Professional*, vol. 19, no. 1, pp. 12–20, Jan.–Feb. 2017.

[8] H. Ye, G. Y. Li, and B.-H. Juang, "Deep reinforcement learning based resource allocation for V2V communications," *IEEE Trans. Veh. Technol.*, vol. 68, no. 4, pp. 3163–3173, Apr. 2019.

[9] L. Zhang, Y. Liang, and D. Niyato, "6G visions: Mobile ultra-broadband, super internet-of-things, and artificial intelligence," *China Commun.*, vol. 16, no. 8, pp. 1–14, Aug. 2019.

[10] M. T. Hammi, B. Hammi, P. Bellot, and A. Serhrouchni, "Bubbles of Trust: A decentralized blockchain-based authentication system for IoT," *Comput. Secur.*, vol. 78, pp. 126–142, Sept. 2018.

[11] Z. Xiong, S. Feng, D. Niyato, P. Wang, and Z. Han, "Cloud/edge computing service management in blockchain networks: Multi-leader multi-follower game-based ADMM for pricing," *IEEE Trans. Serv. Comput.*, vol. 13, no. 2, pp. 356–367, Mar.–Apr. 2020.

[12] T. S. Lopez et al., "Explainable artificial intelligence for wireless communications: A survey," *IEEE Access*, vol. 10, pp. 111234–111250, 2022.

[13] H. Lin, J. Li, and Z. Li, "Blockchain for decentralized spectrum management: Opportunities and challenges," *IEEE Netw.*, vol. 35, no. 1, pp. 320–327, Jan.–Feb. 2021.

[14] I. F. Akyildiz, A. Kak, and S. Nie, "6G and beyond: The future of wireless communications systems," *IEEE Access*, vol. 8, pp. 133995–134030, 2020.
[15] R. Mollah, M. Zhao, and A. K. Azad, "Blockchain for the Internet of Things: Challenges and opportunities," *IEEE Internet Things J.*, vol. 8, no. 4, pp. 2198–2213, Feb. 2021.
[16] Federal Communications Commission (FCC), "Spectrum policy task force report," FCC 02-155, Nov. 2002.
[17] M. Cave, C. Doyle, and W. Webb, *Essentials of Modern Spectrum Management*, Cambridge University Press, 2007.
[18] J. Mitola and G. Q. Maguire, "Cognitive radio: Making software radios more personal," *IEEE Pers. Commun.*, vol. 6, no. 4, pp. 13–18, Aug. 1999.
[19] C. Stevenson et al., "IEEE 802.22: The first cognitive radio wireless regional area network standard," *IEEE Commun. Mag.*, vol. 47, no. 1, pp. 130–138, Jan. 2009.
[20] A. Kliks, P. Kryszkiewicz, and H. Bogucka, "Dynamic spectrum management in 5G networks: From concept to implementation," *IEEE Netw.*, vol. 31, no. 6, pp. 22–29, Dec. 2017.
[21] D. T. Hoang et al., "Reinforcement learning approaches for dynamic spectrum management in next-generation networks," *IEEE Netw.*, vol. 34, no. 6, pp. 268–275, Dec. 2020.
[22] Fuwang Dong, Fan Liu, Yuanhao Cui, etnl , "Sensing as a Service in 6G Perceptive Mobile Networks: Architecture, Advances, and the Road Ahead" *IEEE Netw.*, vol. 38, Issue no 2, pp. 87–96 March 2024.
[23] H. Kim and K. G. Shin, "Efficient discovery of spectrum opportunities with MAC-layer sensing in cognitive radio networks," *IEEE Trans. Mobile Comput.*, vol. 7, no. 5, pp. 533–545, May 2008.
[24] Jon M. Peha , " Sharing Spectrum Through Spectrum Policy Reform and Cognitive Radio" , Proceedings of the IEEE , Volume: 97, Issue: 4, pp. 708– 719 April 2009.
[25] T. Yücek and H. Arslan, "A survey of spectrum sensing algorithms for cognitive radio applications," *IEEE Commun. Surveys Tuts.*, vol. 11, no. 1, pp. 116–130, 2009.
[26] Carla Passiatore, Pietro Camarda, "Inter-Network Resource Sharing in IEEE 802.22 Networks" Communications and Network, vol. 4, no. 2, pp. 111–121, May 2012
[27] ITU Radiocommunication Sector, "Report ITU-R SM.2405: Studies on spectrum management principles for 5G and beyond," Geneva, 2021.
[28] Y. Zeng, Y.-C. Liang, and R. Zhang, "Blindly combining cooperative sensing and predictive modeling for dynamic spectrum access," *IEEE Trans. Veh. Technol.*, vol. 69, no. 9, pp. 10165–10177, Sept. 2020.

[29]. Mohammad Banafaa, Omer Pepeoglu et al.," "A Comprehensive Survey on 5G-and-Beyond Networks With UAVs: Applications, Emerging Technologies, Regulatory Aspects, Research Trends and Challenges," *IEEE Access.*, vol. 12, 2024, pp. 7786–7826, Jan 2024.

[30] Udeme C. Ukpong et al., " Deep reinforcement learning agents for dynamic spectrum access in television whitespace cognitive radio networks," Scientific African 27(3), vol. 27, pp. 1–16, March. 2025.

[31] Yaohua Sun, Mugen Peng,",Application of Machine Learning in Wireless Networks: Key Techniques and Open Issues", IEEE Communications Surveys & Tutorials, Volume 21, Issue 4, pp. 93072–3108, Oct. 2019.

[32] J. Wang et al., "Spectrum sensing based on machine learning techniques for cognitive radio networks," *IEEE Trans. Veh. Technol.*, vol. 68, no. 3, pp. 2483–2493, Mar. 2019.

[33] S. Chen and Q. Yu, "Unsupervised spectrum classification in dynamic wireless environments," *IEEE Access*, vol. 9, pp. 34578–34589, 2021.

[34] Hyebin Park and Yujin Lim, " Deep Reinforcement Learning Based Resource Allocation with Radio Remote Head Grouping and Vehicle Clustering in 5G Vehicular Networks," *Electronics 2021.*, vol. 10, no. 23, pp. 1–19, Dec. 2021.

[35] Y. Liu et al., "Federated learning for 6G communications networks," *IEEE Netw.*, vol. 35, no. 2, pp. 30–36, Mar.–Apr. 2021.

[36] Z. Qin, H. Ye, G. Li, and B. Juang, "Deep learning in physical layer communications," *IEEE Wireless Commun.*, vol. 26, no. 2, pp. 93–99, Apr. 2019.

[37] M. A. Alhussein et al., "Spectrum data augmentation using GANs for deep-learning-based wireless signal classification," *IEEE Access*, vol. 8, pp. 214480–214490, 2020.

[38] F. Tang, Y. Kawamoto, and N. Kato, "Future intelligent and secure vehicular network toward 6G: Multi-agent and blockchain perspectives," *IEEE Wireless Commun.*, vol. 29, no. 1, pp. 84–91, Feb. 2022

[39] C. Jiang, H. Zhang, Y. Ren, and M. Debbah, "Machine learning paradigms for next-generation wireless networks," *IEEE Wireless Commun.*, vol. 24, no. 2, pp. 98–105, Apr. 2017.

[40] S. Chen, R. Li, and Z. Zhao, "AI-enabled intelligent wireless networks: From cognitive radios to 6G," *IEEE Trans. Wireless Commun.*, vol. 27, no. 5, pp. 10–19, Oct. 2020.

[41] F. Tang et al., "AI for proactive resource management in 6G networks," *IEEE Commun. Mag.*, vol. 58, no. 10, pp. 76–82, Oct. 2020.

[42] J. Wang, L. Zhao, and H. Xu, "Hybrid machine learning framework for efficient spectrum classification," *IEEE Access*, vol. 9, pp. 32108–32119, 2021.

[43] Rajilal Manathala Vijayan, Fabrizio Granelli , A. Umamakeswari " An Adaptive Framework for Resource Allocation Management in 5G Vehicular Networks ,". International Journal of Interactive Multimedia and Artificial Intelligence, Vol. 9, no.5 , pp. 141–153, Apr. 2025.

[44] Zhaohui Yang,, Mingzhe Chen et al., " Federated Learning for 6G: Applications, Challenges, and Opportunities," , *Elsevier Engineering* , vol. 8, pp. 33–41, Jan. 2022.

[45] M. A. Ferrag et al., "Blockchain technologies for the internet of things: Research issues and challenges," *IEEE Internet Things J.*, vol. 9, no. 1, pp. 569–595, Jan. 2022.

[46] R. Xu, L. Chen, and Z. Li, "Decentralized dynamic spectrum trading using blockchain and smart contracts," *IEEE Access*, vol. 8, pp. 185120–185131, 2020.

[47] A. S. Al-Fuqaha and N. S. Al-Karaki, "Blockchain for spectrum sharing and trading in cognitive radio networks," *IEEE Commun. Surveys Tuts.*, vol. 23, no. 3, pp. 1680–1704, 2021.

[48] J. Gao, L. Xu, and Y. Zhang, "Blockchain-assisted trustworthy AI for 6G communications," *IEEE Netw.*, vol. 36, no. 1, pp. 109–115, Jan.–Feb. 2022.

[49] A. Kliks, F. Paisana, and M. Matinmikko-Blue, "Spectrum management for future wireless networks: Challenges and opportunities," *IEEE Commun. Mag.*, vol. 58, no. 6, pp. 66–72, Jun. 2020.

[50] M. A. Brown, A. Islam, and R. Tafazolli, "Regulatory evolution for dynamic spectrum access: A roadmap toward 6G," *IEEE Access*, vol. 9, pp. 118230–118245, 2021.

[51] ITU-R, Report ITU-R SM.2400-2, Spectrum management principles for intelligent radio systems, Geneva, 2022.

CHAPTER 5

Artificial Intelligence Solutions to Dynamic Spectrum Prediction and Optimization in 5G and 6G Systems

Swati Jaiswal, Spandan Surdas, Renu Kachhoria, and Amitesh Jaiswal

5.1 INTRODUCTION

The radio frequency (RF) spectrum has been further strained by the explosive growth in the number of wireless gadgets and the data-intensive applications. The new generation of communication systems such as 5G and the next generation 6G will have the capacity to support different requirements such as enhanced mobile broadband (eMBB), ultra-reliable low-latency communication (URLLC), and massive machine-type communication (mMTC). But the existing policy of spectrum allocation in these dynamic and heterogeneous environments is inefficient, and results in the underutilization and congestion of spectrum (Akyildiz et al., 2006).

The concept of cognitive radio (CR) and dynamic spectrum access (DSA) had to emerge in an attempt to make the spectrum more efficient through opportunistic access of the unlicensed users to underutilized licensed bands without causing interference to the primary users (Mitola and Maguire, 1999). The key components of these paradigms are spectrum sensing and spectrum prediction. Classical prediction techniques,

DOI: 10.1201/9781003731597-5

including autoregressive models, hidden Markov models (HMMs), and other statistical time-series techniques, have been popularly investigated but are unable to reproduce the highly dynamic and non-stationary nature of real-world spectrum usage (Sun et al., 2013).

New developments in artificial intelligence (AI) and machine learning (ML) offer some hopeful alternatives. Deep learning (DL) approaches that have shown great improvements in predicting spectrum occupancy via the learning of complex spatial and temporal interactions include convolutional neural networks (CNNs) and recurrent neural networks (RNNs) (Wang et al., 2020). Likewise, reinforcement learning (RL) and deep reinforcement learning (DRL) algorithms have been effectively used to DSA, power management, and interference management and provide adaptive and autonomous decision-making abilities (Xu et al., 2012; Zhang et al., 2019). Intelligent spectrum management is based on the integration of prediction and optimization. Spectrum prediction gives the ability to predict the future availability and optimization is a proactive allocation of resources to achieve maximum spectral efficiency and minimum interference (Zhao et al., 2021). It is especially applicable to applications like Internet of Things (IoT), vehicular networks (V2X), satellite-terrestrial communication, and so on, where conditions are dynamic and quality-of-service (QoS) requirements are extremely strict (Chen et al., 2021).

However, there are still problems. The lack of data, non-stationarity, and adversarial attacks can potentially degrade the accuracy of a prediction. Limits on the use of complex models are imposed by resource-constrained edge devices, and further restrictions are imposed by explainability and compliance with regulations (Sagduyu et al., 2019; Letaief et al., 2019). These matters are important to address in order to deploy in the real world. Research, including works by Pan et al. (2024), has pointed to the transformative impact of DL models in spectrum management, which deal with non-stationary signals, time dependencies of signals and spatial dependencies. On the same note, a novel big data-based model that was created by Zhang et al. (2023) presented the graph-based predictive analysis of spectrum resources as an example of utilizing large-scale data to achieve better decision-making. Deep sequence learning combined with temporal attention tasks with spectral feature mining, as observed in the Bi-ConvLSTM and Seq2Seq networks as proposed by Li et al. (2024), can be considered the state-of-the-art of CR networks as preconditioning the adaptive and intelligent spectrum management.

Recent breakthroughs in DL have made a great impact on enhancing the ability of spectrum prediction and optimization in dynamic wireless environments. Unlike conventional machine learning methods, deep neural networks are able to automatically extract hierarchical features from a large-scale spectrum data sets and capture complex temporal-spectral dependencies in RF signals. Architectures, such as CNNs and RNNs, have shown great performance in models of spectrum occupancy patterns and channel availability prediction in CR networks (Wang et al., 2020; Chen et al., 2020). More recent progress involves the development of transformer inspired and hybrid DL architectures inclusive of attention mechanisms and sequence learning ability to manage long-term dependencies and non-stationary spectrum behavior (Pan et al., 2024; Li et al., 2024). In addition to prediction, DL techniques are also incorporated in RL frameworks with greater integration to address adaptive spectrum allocation and interference management in the next generation networks (Zhao et al., 2021; Zhang et al., 2019). These developments illustrate DL has emerged as an important enabling technology for intelligent and autonomous spectrum management for future wireless communication systems.

This chapter gives an exhaustive overview of spectrum prediction and optimization techniques for next-generation wireless networks using AI/ML techniques. Section 2 presents the spectrum prediction models and methods, including traditional statistical methods and DL methods. Section 3 introduces spectrum optimization and resource allocation techniques, including the conventional methods and the methods based on AI. Section 4 describes the DL for interference mitigation for intelligent spectrum management. Section 5 examines some of the important applications in next-generation networks, and Section 6 discusses challenges and future research directions. Finally, some practical considerations of implementation and design insights are summarized in the final section.

5.2 SPECTRUM PREDICTION: MODELS AND METHODS

5.2.1 Problem Formulation

Spectrum prediction is an important component regarding DSA because it predicts the availability of frequency bands. The traditional spectrum allocation framework based on fixed licensing is lack of dynamics and non-homogenous spectrum allocation system to 5G and beyond. The need to have spectrum awareness to enable opportunistic communication was stated in early frameworks of CR, which was proposed (Mitola and Maguire, 1999). Follow-up surveys, including the one conducted (Akyildiz et al., 2006), that described that DSA can be of great benefit to improve congested environment utilization.

Spectrum prediction is usually the learning of patterns of RF use, both spatially and in time. Markov chains and autoregressive techniques became popular as statistical modeling methods in the early days, but were found to be poor predictors in high traffic variability. With the advent of AI/ML, DL approaches have been found to be more capable. Indicatively, Wang et al. (2020), used deep neural networks to perform phase retrieval to optimize spectrum sensing, whereas the other author used RNNs to forecast temporal occupancy, which is better than the performance of classical predictors.

Recent works have moved to the context-informed prediction that entails the utilization of multifarious heterogeneous information sources including cellular networks, sensors, and IoT devices to improve the accuracy of the forecast. Elsayed et al. (2019), pointed out that spectrum prediction with the help of AI may provide near real-time decision-making in 6G, especially when working with extremely dynamic vehicle and satellite networks. These developments indicate the shift in the reactive to proactive spectrum management capability and this has opened up increased levels of efficiency and reliability in wireless systems.

x_t, f can be defined as the observed state (or measurement) on frequency band f at time t. A predictive model seeks:

$$\hat{x}_t + \tau, f = M\left(\left\{x_{t-k+1} \quad x_t, f\right\}, content\right)$$

where τ is the prediction horizon, and "context" may include spatial features (neighboring bands, location), network load, device mobility, etc.

Costs on prediction errors are of two types in CR situations:

- False negative (predict free but occupied in fact) – interference to primary user.
- False positive (predict occupied and free) – lost opportunity (decreased spectral efficiency)

In that way, prediction models have to trade accuracy, latency, and robustness.

Spectrum prediction aims at performing predictions on the future channel occupation patterns and utilize them by the secondary users on the opportunistic access to unused frequency bands. The classical methods were based on statistical and stochastic models Markov chains, autoregressive (AR) models, and hidden Markov models (HMMs). These methods are computationally efficient but cannot deal with non-linear and high-dimensional relations that occur in the contemporary wireless environment. The relative comparison of Liu et al. (2023) has shown that classical models usually do not work when faced with the complexity of signal variations due to multi-user interference and time-varying conditions.

The most recent trends of 2023–2025 are the clear transition to data-driven spectrum prediction. Pan et al. (2024) came up with a Deep 3D Pyramid Vision Transformer model that was able to capture spatial-temporal-spectral dependencies at the same time, which significantly outperformed LSTM-based baselines. In the same vein, Li et al. (2024) suggested the use of multi-channel multi-step models, which used hybrid Transformer-BiLSTM models to manage long-term associations in many channels. In the meantime, Wen et al. (2023) presented a graph neural network (GNN) model that simultaneously with space and time captures the radio signal, which allows predictive performance in diverse and varying frequency landscapes. All these developments have increased the accuracy and strength of prediction, so models using AI will become the foundation of cognitive communication systems of the future.

5.2.2 Traditional and Statistical Prediction Techniques

The initial spectrum prediction studies were based on statistical models that represent time-varying correlation in spectrum occupancy. These models are based on the assumption that spectrum occupancy is probabilistically distributed, which is frequently based on a measurement of traffic or past channel usage.

- Markov Models: Markov Modeling Spectrum occupancy is often represented as a Markov process, in which the state of the system only depends on the last state. The unobserved channel states were inferred on the basis of noisy observations by means of Hidden Markov models (HMMs) and allowed the prediction of idle and occupied slots.
- Autoregressive Models (AR/ARIMA): ARIMA forecasts occupancy in the future using the trends of the past. The strengths of these

models are that they are good when the traffic is stationary but they do not work well in highly dynamic situations (Sun et al., 2013).

- Bayesian Inference: Bayesian models are used to estimate the probabilities of spectrum availability on the basis of previous distributions. They provide a principled approach to take uncertainty into consideration, but the computational costs grow as the network size grows.

Statistical methods are computationally efficient but do not reflect nonlinearities and context dependencies in the modern wireless traffic. This drawback inspired the use of ML based methods.

5.2.3 Deep Learning (DL) Approaches

DL has transformed spectrum prediction, in which the learning of spatial-temporal features on raw sensing data is performed automatically. DL models are data-scaling and environmentally adaptable to a complex wireless environment.

- RNNs: RNNs with their conceivable variations (LSTMs, GRUs) discover the temporal correlations in the spectrum data, which makes them capable of predicting long-term occupancy accurately.
- CNNs: CNNs are methods that learn spatial patterns in frequency and time using spectrum data in the form of images (spectrograms). This method enhances the strength when the conditions are noisy.
- DRL: DRL is a hybrid of prediction and decision-making that allows the agents to anticipate occupancy and at the same time combined with access strategies optimization.
- Hybrid Models: The new models are based on CNNs and RNNs with joint learning in space-temporal, which has reached the state of the art in prediction accuracy. Elsayed and Erol-Kantarci have noted hybrid models as being pivotal to 6G networks of the future.

In contrast to ML, the DL techniques automatically extract features, and thus, manual work is minimized, as provided in Table 5.1. They, however, need large datasets and lots of computational power and are overfitting unless well-constructed. The most important breakthrough in intelligent spectrum management is the DL-based spectrum prediction. The Deep

TABLE 5.1 Comparative Analysis of Statistical, ML and Deep Learning Approaches

Category	Example Methods	Strengths	Limitations	Key Reference
Statistical	Markov models, ARIMA, Bayesian inference	Simple, interpretable, low complexity	Poor accuracy in dynamic traffic, limited adaptability	Akyildiz et al. (2006); Sun et al. (2013)
Machine Learning	SVM, KNN, Decision Trees, Random Forests	Better than statistics in nonlinear settings, adaptable	Needs feature engineering, scalability issues	Xu et al. (2012); Zhang et al. (2019); Zhao et al. (2021); Chen et al. (2021)
Deep Learning	RNN, CNN, DRL, Hybrid CNN-RNN	High accuracy, automatic feature learning, robust	Data hungry, computationally intensive, "black box"	Wang et al. (2020) Xu et al. (2012); Zhang et al. (2019); Sagduyu et al. (2019)

Pyramid Vision Transformer proposed by Pan et al. (2024) has the ability to learn a combination of hierarchical time and spectral features and is capable of achieving over 95% accuracy when it comes to multi-band prediction. On a different experiment, Pan et al. (2024) have shown that the combination of Transformer and Bi-LSTM layers can be very effective to come up with long range channel dependencies that are superior to traditional LSTM and GRU models. The hybrid model proposed by Guo et al. (2023) to fuse spectral and spatial relationships in dynamic spectrum settings is an important move towards predictive generalization. These DL methods have been tested with real-world data and it demonstrates strong generalization to SNR variations, channel dynamics and interference. The 2023–2025 trend is evidently focused on hybridization, which is the combination of CNNs, LSTMs, GNNs Zhu et al. (2025), and Transformers, to improve the spatiotemporal learning process and reduce the latency of the prediction process.

5.3 SPECTRUM OPTIMIZATION AND RESOURCE ALLOCATION

Prediction on its own is not enough without optimization strategies to be able to allocate spectrum efficiently between competing users. Optimization aims at maximizing the total system throughput and being fair and reducing interference, as depicted in Figure 5.1.

Spectrum sharing has traditionally been handled using game-theoretics. Xu and others proved that global optimization was possible to be reached in opportunistic access through local interaction games. RL has emerged as a popular method with the emergence of ML. Zhao et al. (2021), discussed that DRL allows agents to learn adaptive allocation strategies by engaging with the surrounding environment and this makes it highly applicable in non-stationary wireless environments.

Ye et al. (2019) suggested a DRL-based framework of vehicle-to-vehicle (V2V) communications to allocate spectrum resources dynamically to satisfy the low-latency demands. On the same note, Bi et al. (2021) proposed a joint spectrum and power optimization framework based on DL, and it showed a high level of energy efficiency. Chen et al. (2021) also extended this line of direction by applying federated learning (FL), which allows distributed devices to jointly optimize the use of the spectrum without sharing raw information. Spectrum prediction and optimization is the integration of both the data-driven intelligence and adaptive decision-making to manage the spectrum efficiently. Figure 5.2 depicts how machine learning and DL models process historical and live spectrum data to forecast the future availability of the spectrum, such as the occupancy distributions, interference, and network environments.

The optimization engines are then employed to make optimal allocation choices using these predictions through RL, DRL, FL, and game-theoretic methods. The resultant framework facilitates bandwidth management, power management, as well as channel access to support applications like IoT, V2V, 6G systems, and immersive VR/AR environments. Moreover, prediction results may also be directly used by applications to proactively change in response to predicted spectrum conditions.

Other than prediction, optimization is more important in the management of the spectrum resources among competing users. The conventional optimization methods, including the linear programming and the heuristic methods, are frequently not sufficient to manage the de facto dynamic and high-dimensional systems. Therefore, there is an increased popularity of AI-based optimization techniques. Liu et al. (2023) demonstrated dynamically spectrum access technique in dynamic conditions of sensing with the help of DRL, and Chen et al. (2024) proposed the idea of a distributed dueling double DQN system that would allow to better utilize the channel

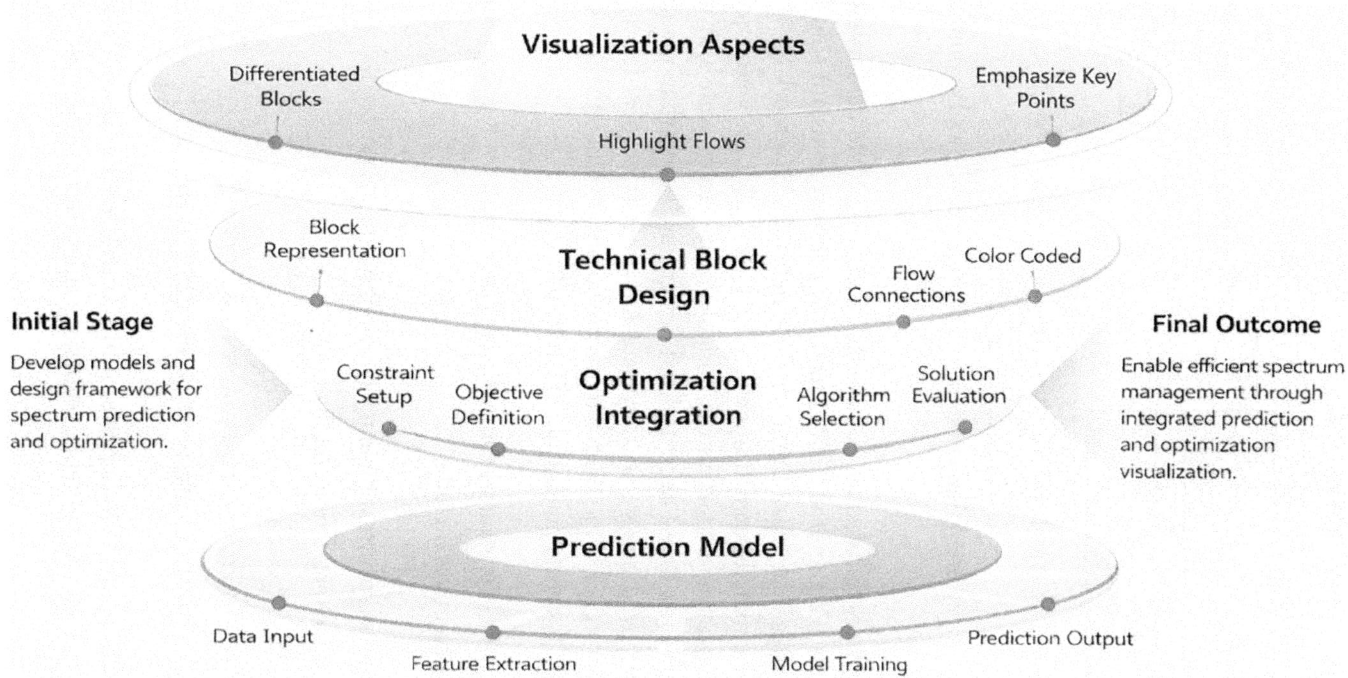

FIGURE 5.1 Visualization aspects of spectrum prediction model.

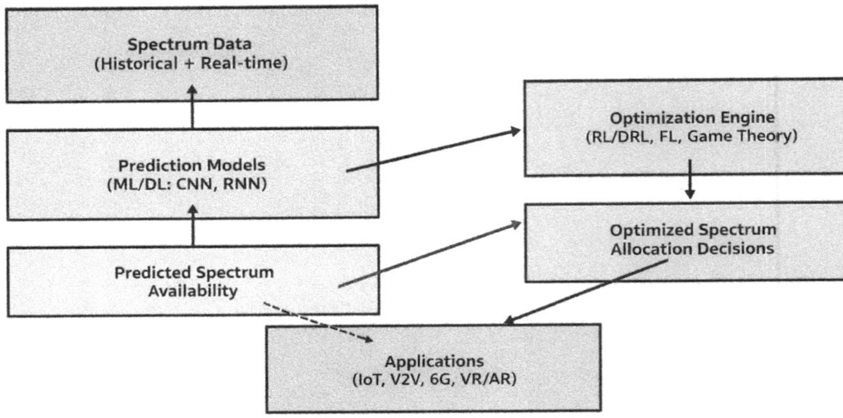

FIGURE 5.2 Integration framework of spectrum prediction and optimization.

by several users and reduce collisions. In addition, federated multi-agent RL methods (Zhang et al., 2023) have been shown as useful solutions to heterogeneous internet of things and car networks. The work of Pan et al. (2024) presents federated frameworks of DRL based on which collaboration in allocating a spectrum is possible with greater scale and without violating privacy. These decentralized designs are better than the conventional centralized designs as they gain the benefits of local learning and global policy aggregation, which to a considerable extent enhances spectral efficiency in the next-generation 6G networks.

5.3.1 Problem Types and Objectives

Common objectives include:

- Maximize spectral efficiency (throughput per Hz)
- Minimize interference/collisions
- Fairness across users
- Energy efficiency (reduce power consumption)
- Latency constraints (especially in URLLC scenarios)

Constraints may include power limits, interference thresholds, quality-of-service (QoS) requirements, regulatory rules, and user priorities.

5.3.2 Conventional Optimization Methods

Before the emergence of AI, game-theoretic and mathematical programming dominated spectrum optimization. These models are based on knowledge of system parameters, and seek to optimize the throughput or fairness given constraints.

- Linear and Nonlinear Programming: an optimization models applied both linear and nonlinear (convex) programming to assign spectrum to the users with the least amount of interference. They are efficient to small-scale problems, but fail to scale to complex wireless environments.
- Game Theory: Game-theoretic models represent the competition of users to the spectrum. It was shown that the approach could result in global optimization of opportunistic spectrum access in the form of local interaction games.
- Nash Equilibrium Notions: The Nash equilibrium notions were extensively employed though tended to need perfect information and centralization.
- Heuristic Algorithms: Metaheuristics, including genetic algorithms (GA) and particle swarm optimization (PSO), were used in spectrum allocation. The methods work well in large solution spaces, but can have convergence problems.
- Dynamic Programming: Dynamic programming techniques were used to solve spectrum allocation problems by breaking problems into subproblems. In real-time, however, they are computationally very costly.

Traditional optimization techniques offered a good theoretical base though they did not perform well in the dynamic, uncertain, and large-scale wireless environment, which led to the move to AI-based techniques. The synergies between model-based and AI-based optimization have also been highlighted by Zappone et al. (2019).

5.3.3 AI/ML-Based Optimization Techniques

Adaptive, data-driven optimization methods that were presented in AI and ML operate better in dynamic wireless conditions compared to traditional algorithms. Figure 5.3 of AI/ML-based spectrum optimization methods

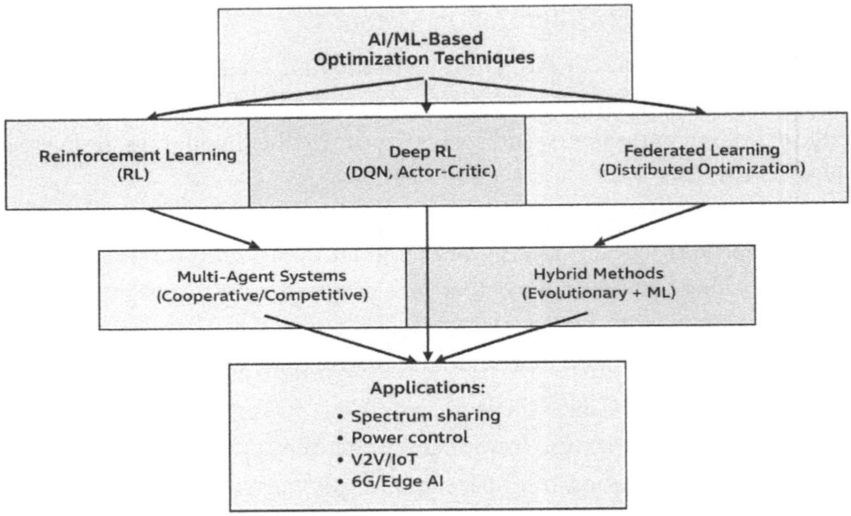

FIGURE 5.3 AI/ML-based spectrum optimization techniques.

depicts the use of AI in a dynamic and efficient distribution of resources within a wireless system. The fundamental AI/ML optimization can be further divided into various methods, all of which approach spectrum issues in a unique way. RL techniques allow agents to learn allocation algorithms by learning via trial and error interaction with the environment, and DRL generalizes this ability by applying neural networks to high-dimensional and complex spectrum states. FL plays its role by allowing distributed optimization among a number of nodes without the need to collect the data centrally, thereby enhancing privacy and scalability. To supplement these, multi-agent systems are used to model cooperative and competitive interactions among the different users or networks and are therefore appropriate in the heterogeneous spectrum environment. Hybrid techniques are a combination of evolutionary algorithms, heuristics, or optimization theory and machine learning to obtain the performance of strong adaptation and robustness. Taken altogether, all these methods result in the practical application of spectrum sharing, power management, dynamic channel access, and resource allocation in the context of IoT, V2V communications, and new 6G and edge AI environments. The hierarchical composition of the diagram illustrates the contribution of each technique toward the development of intelligent, scalable, and adaptable spectrum optimization solutions to next-generation networks.

- RL: RL enables the agents to optimize the spectrum allocation by means of interacting with the environment. Zhao et al. (2021) overviewed DRL applications and discovered that RL allows the adaptive strategies with no system modeling.
- DRL: Ye and Li (2018) utilized DRL to V2V communications, where the spectrum allocation was performed in real-time and with a very high latency threshold. Bi et al. (2021), suggested joint optimization of the allocation of spectrum and power by using DRL to enhance the spectral performance and fairness.
- FL: FL shares learning devices to do collaborative optimization without centralized information. Chen et al. (2020) have shown a collaborative learning-communication system to federated spectrum optimization to enhance the privacy and scalability (Qamar et al., 2024).
- Multi-Agent Systems: In heterogeneous networks, the spectrum access is coordinated by several RL agents. He et al. (2022) put an accent on integrated optimization in 6G where multi-agent learning will balance between the spectrum, power, and the computing resources.
- Evolutionary and Swarm Intelligence Techniques: In the recent times, hybrids are used in the form of ML combined with GA or PSO to achieve scalability over the spectrum. They are not as popular as DRL, but they are applicable in situations where large solution spaces have to be explored.

The true power of AI/ML technologies is based on the fact that these technologies are able to scale to dense networks, adapt to the presence of uncertainties, and combine with real-time prediction and are the key to spectrum management in 5G and future 6G technology. The comparison of conventional and AI/ML-based spectrum optimization (Figure 5.4) indicates the transition of non-adaptive and assumption-based approaches to adaptive and data-driven intelligence. Conventional methods employ mathematical programming, allusion, and game-theory models, which usually presuppose perfect information about the network and cannot embrace dynamics, scalability, and real-time decision-making. Quite to the contrary, AI/ML-based approaches, i.e., RL, DRL, FL, and multi-agent systems, enable to allocate spectrums adaptively based on the dynamically

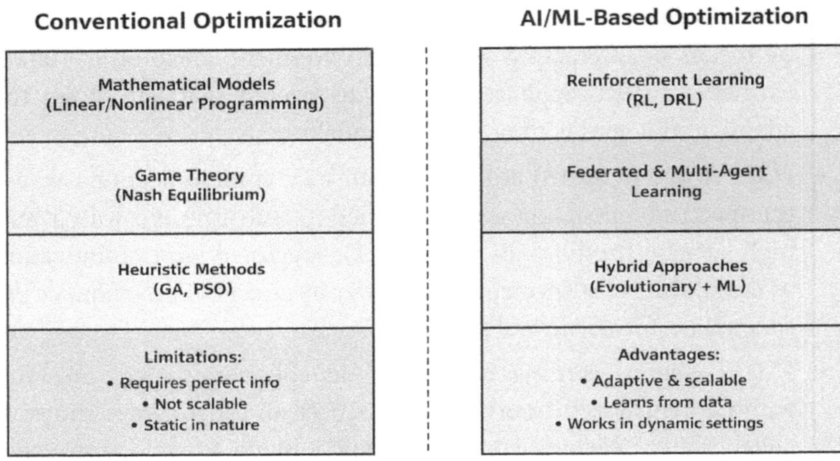

FIGURE 5.4 Comparison: Conventional vs AI/ML-based spectrum optimization.

shifting environment, incomplete or noisy information. Such transition enhances flexibility, scalability, and robustness, which, in the final, will allow more effective spectrum sharing, stronger control of power, and more support of next-generation services such as IoT, V2V, VR/AR, and 6G networks.

One of the prospects of the spectrum management field is the integration of traditional and AI/ML-based optimization techniques that take the advantages of both paradigms and merge them as provided in Figure 5.5. Traditional mathematical programming and heuristic models offer good theoretical guarantees, interpretability, and convergence properties. They are however usually hard and computationally expensive in large scale or highly dynamic applications. Conversely, AI/ML-based methods that include RL, deep neural models, and federated optimization introduce the flexibility, scalability, and the capacity to learn directly on data, but they might not provide any transparency and can consume a lot of training resources.

A combination of such strategies can result in more stability and more carefully designed frameworks, such as applying ML-based prediction to propose improved inputs to mathematical solvers, or putting optimization constraints on top of RL models to verify feasibility and fault-tolerance. These hybrid solutions are specifically well applied to the real world application when spectrum allocation has to be able to balance the theoretical efficiency, practical flexibility and regulatory compliance.

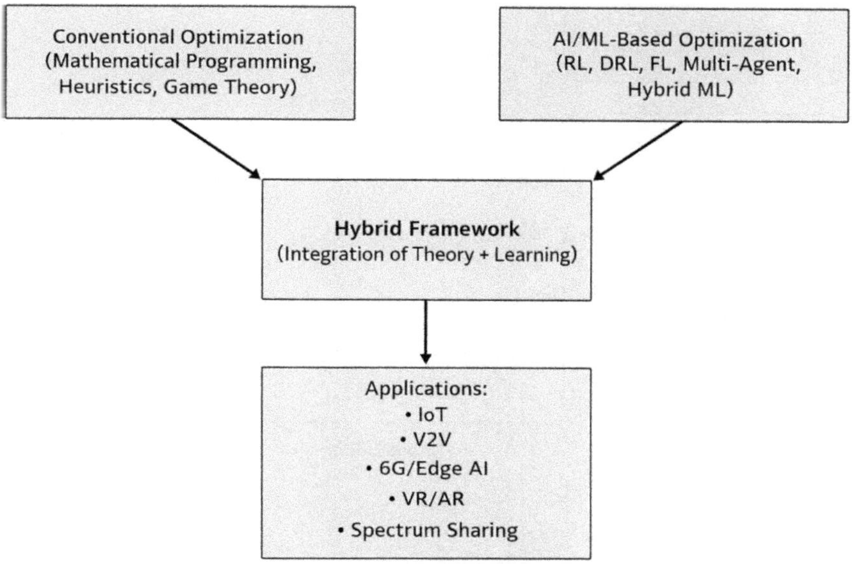

FIGURE 5.5 Hybridization of traditional with AI/ML-based spectrum optimization.

This synergy would allow the design of stronger, interpretive, and adaptive strategies of spectrum optimization to the next generation wireless networks.

5.3.4 Integration: Prediction + Optimization

Spectrum prediction and optimization co-location has got a lot of advantages as it provides intelligent and proactive control of spectrum. Prediction gives information about the availability of spectrum in future, whereas optimization determines resources to be allocated so that it is efficient. Figure 5.6 demonstrates that the integration framework starts with the spectrum data retrieved in historical records and real-time sensing that should be processed with the help of feature extraction and preprocessing to enable efficient analysis. They are then predicted by using machine learning and DL models, including LSTM, CNNs, and Transformers, to predict spectrum occupancy. Optimization engines that utilize such predictions apply a variety of methods, such as RL (RL/DRL), FL, and game-theoretic models to identify the most desirable strategy to allocate resources, such as channel assignment, power balance, and time scheduling. The optimized decisions obtained serve other applications like IoT, V2V communications,

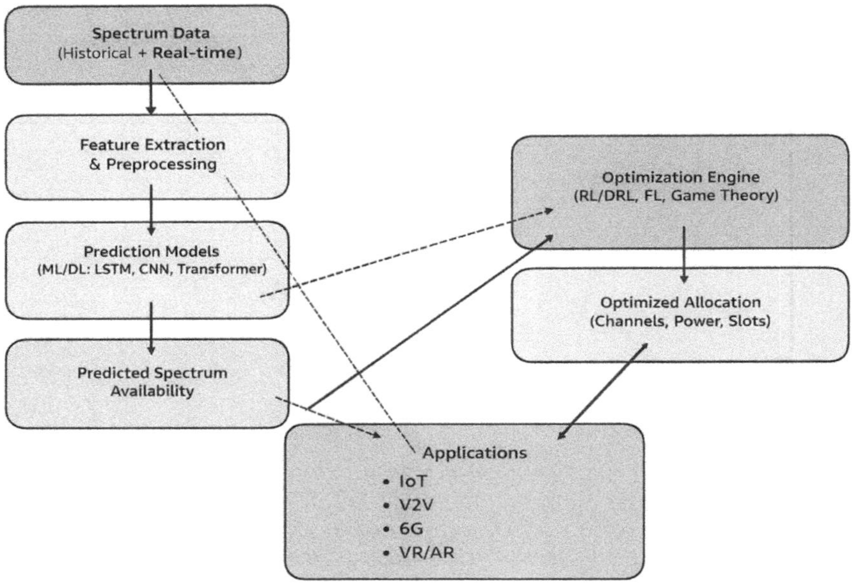

FIGURE 5.6 Integration of prediction with optimization.

6G networks, and VR/AR systems by assuring effective and reliable use of the spectra. Moreover, prediction results could enable proactive adaptation on the application level directly and a feedback loop between applications and data layer is the key to ongoing learning and adaptability in dynamic wireless environment.

5.3.4.1 Predict-then-Optimize Frameworks:

Access strategy predictive models are trained with DL and then optimized with RL to make predictions of spectrum occupancy. Chen et al. (2020) and Wang et al. (2020) demonstrated the direct improvement of allocation efficiency by the prediction accuracy.

5.3.4.2 Collaborative Learning:

In some frameworks, they combine both the task of prediction and optimization by performing the tasks jointly. Ye and Li (2018) came up with a DRL model, which predicts both spectrum states and resources allocation at the same time in V2V communications.

5.3.4.3 Context-Aware Optimization:

It focused on hybrid AI models that combine prediction (through CNN/RNN) and optimization (through RL) to 6G systems, specifically in highly mobile networks such as vehicular and satellite networks.

5.3.4.4 Federated Prediction-Optimization:

Another approach came up with FL in distributed prediction and optimization to allow devices to collaborate with privacy protection. This is essential in the IoT and edge-driven 6G.

5.3.4.5 Secure and Robust Integration:

Lu et al. (2021) examined blockchain and FL as the means to provide secure prediction-optimization integration, which will prevent the risks of adversarial manipulation.

This integration provides a paradigm shift of the reactive spectrum usage to an intelligent, proactive, and autonomous spectrum use, leading to the next-generation networks. Table 5.2 provides a comparative analysis of various methods used for optimization techniques.

TABLE 5.2 Comparative Analysis of Various Optimization Techniques

Category	Example Methods	Strengths	Limitations	Key Reference
Conventional Optimization	Linear programming, Game theory, DP, GA, PSO	Strong theoretical foundation, interpretable	Poor scalability, assumes perfect info	Zhao et al. (2021)
AI/ML-Based	RL, DRL, FL,	Adaptive,	Requires large	Chen et al. (2021); Sagduyu et al. (2019); Letaief et al. (2019); Chen et al. (2020); Elsayed and Erol-Kantarci (2019)
Optimization	Multi-agent learning	scalable, works in dynamic settings	training data, stability issues	
Integrated Prediction + Optimization	Predict-then-optimize, DRL-based joint models, FL with prediction	Proactive, intelligent, robust	High complexity, data privacy/ security issues	Wang et al. (2020) Xu et al. (2012); Zhang et al. (2019); Sagduyu et al. (2019); Letaief et al. (2019); Zorzi et al. (2017); Pan et al. (2024)

It is projected that the future 6G networks will incorporate optimization of various resources- spectrum, energy, and computing on a common platform. He et al. (2022) have addressed the paper on integrated resource allocation in 6G, in which AIs are important to balance the heterogeneous needs. On a larger scale, Arulkumaran et al. (2017) gave a review of DRL in resource allocation, making it a groundbreaking optimization instrument in wireless resource allocation.

5.4 DL FOR INTERFERENCE MITIGATION

DL has become a formidable paired approach to interference mitigation in modern wireless communication systems by allowing intelligent interference detection, classification, cancellation, and interference prediction. Unlike traditional signal processing techniques that use predefined analytical models, DL techniques do not require predefined analytical models, but can be used to learn complex relationships between spectra and time directly from the data, making them suitable for dynamic and heterogeneous wireless environments. Recent investigation has proven that deep neural networks greatly enhance the accuracy of interference detection, the reliability of channel prediction, and the reliability of signal reconstruction under a low signal-to-noise ratio (SNR) environment (O'Shea and Hoydis, 2017; Qin et al., 2019). Furthermore, DL together with adaptive spectrum management techniques can provide proactive interference mitigation in 5G networks and future 6G networks (Letaief et al., 2019; Zhang et al., 2019).

5.4.1 DL in Interference Detection and Classification

DL techniques have also shown a significant improvement in the detection and classification of the interference in CR and DSA systems. CNN-based models are good at time-frequency spectrogram feature analysis to obtain the spatial features, which makes it appropriate to distinguish between different types of interference using multi-classes classification, e.g., co-channel, adjacent-channel, and impulsive noise. Moreover, the RNN and LSTM models are capable of capturing time dynamics enabling them to effectively track and predict how time-varying interference changes. The methods attain classification rates of 90–95 percent at moderate SNRs and perform stronger and more versatile to changing wireless environments compared to conventional algorithms, such as matched filtering and energy detection (Mumtaz et al., 2016; O'Shea and Hoydis, 2017; Qin et al., 2019; Wang et al., 2020).

5.4.2 Interference Cancellation Based on Neural Network

Neural network-based interference cancellation methods help address the issue of communication reliability by suppressing unwanted signals and achieving a better quality of the desired communications. Autoencoder-based architectures are widely applied for signal denoising where the neural network learns to represent signal in a compact manner and reconstruct the signal of interest removing the interference elements. More advanced techniques make use of generative adversarial networks (GANs) to model interference characteristics and suppress them using adversarial learning. Another promising approach is to use deep filtering networks which learn good filtering functions from data itself and are in many cases better than the classical filtering methods of Wiener and Kalman. These neural network based interference cancellation approaches usually achieve signal to interference ratios (SIR) improvements of about 8–15 dB and large bit error rate (BER) reductions, and have shown an improvement in performance when working in moderate and strong interference environments (Elsayed et al., 2021; Jagannath et al., 2021; Lu et al., 2021; Qin et al., 2019).

5.4.3 Advanced Channel Prediction

DL-based channel prediction for proactive interference mitigation and adaptive resource allocation for next-generation wireless networks. Long short-term memory (LSTM) and gated recurrent unit (GRU) networks are popular choices for long-term memory (LTM) because they are suitable for time series prediction of channel quality indicators like signal-to-noise ratio (SNR), channel gain, and interference levels due to their ability to capture temporal correlations in wireless channels. For wideband and multichannel systems transformer-based models can offer better performance since they can learn long range dependency over frequency bands and time intervals. Hybrid architectures have been used to further improve the prediction accuracy by joint modeling of the spatial and temporal channel characteristics using CNNs and LSTM networks. Comparative studies show that the DL models can provide much better performance than the classical statistical methods like ARIMA in the dynamic wireless environments, with transformer-based models typically showing the best prediction accuracy followed by the LSTM-based models (Chen et al., 2020; Qin et al., 2019; Zhang et al., 2019).

5.4.4 Transfer Learning and Implementing

Transfer learning has become an efficacious approach in practical implementation of interference mitigation systems based on DL. Model training can occur by first training the model with large simulated data sets, and then by fine-tuning with small data sets from actual measurements, which can cut data requirements by about one to two orders of magnitude. This approach is especially useful in the context of wireless environments, in which dataset of labeled spectrum is limited. Practical deployment must also take hardware and latency constraints into account, which is especially true for edge devices and IoT systems with limited computational resources. Techniques such as model quantization, pruning, and lightweight neural network architectures are being used to reduce the computational complexity and memory requirements and maintain an acceptable performance. These optimization techniques make implementation of DL-based interference mitigation of next-generation wireless networks (Elsayed et al., 2021; Jagannath et al., 2021; Letaief et al., 2019) possible in real time.

5.5 APPLICATIONS IN NEXT-GENERATION NETWORKS

The new 5G and 6G networks have made AI/ML-based spectrum prediction and optimization core in order to have adaptive, flexible, and intelligent wireless systems. As emphasized by Letaief et al. (2019), one of the most important types of AI-based spectrum management is the ability to self-organize and adapt a network in real time, which enables 6G networks to operate. In the same way, Zhang et al. (2019) showed that DL assists in dynamic resource allocation, interference reduction, and mobility forecasting of current networks. The latest developments have enhanced more the use of AI optimization of spectrums in high-throughput and low-latency like VR, AR, and XR. According to the studies of Pan et al. (2024), the combination of DL-based prediction and RL allows efficiently managing dynamic spectrum conditions and enhancing QoS, throughput and latency. The intelligent transportation systems and vehicle spectrum management are also critical in AI (Deshpande et al., 2024). According to Liu et al. (2023) and Chen et al. (2024), adaptive channel and power allocation, realized with the help of DRL-based frameworks, can make the process more reliable and mitigate collisions in congested traffic. The risk of adversarial machine learning decreases predictive capacity, and the existing paradigms of distributed and collaborative intelligence are required to improve resilience (Zhu et al., 2025; Zorzi et al., 2017). In the future, 6G networks are likely to have spectrum prediction and optimization networks built in on a multi-layer of terrestrial, aerial, and satellite networks to enable URLLC, massive IoT and space-air-ground communication (Dang et al., 2020). These developments together with FL-driven optimization are introducing self-aware, scalable and energy-efficient wireless ecosystems with the potential to support the next-generation applications like autonomous systems, smart healthcare and immersive digital services.

5.6 CHALLENGES AND FUTURE DIRECTIONS

Even though the future of AI/ML-based spectrum management as a promise is high, numerous issues need to be resolved until large-scale implementation can happen. One of them is a trade-off that exists between model and AI designs where high-performing DL and RL models can be not easily interpreted and transparent, making it impossible to trust and effectively implement them in practice (Ukpong et al., 2025). Other issues, as pointed out by Sabir et al. (2024), are high computational costs, the unavailability of real-life examples, and benchmarking in AI-enabled spectrum management in 6G (Liu et al., 2021). Moreover, the computational complexity and latency are still significant obstacles because the models that require many resources are not easily implemented in edge and IoT devices with small hardware and energy constraints. Problems like convergence, stability in an RL-based approach, particularly in non-stationary and multi-agent settings, aggravate real-time implementation even further. The explainability, security and robustness of AI-driven decisions is another important issue of concern. Although explainable AI (XAI) attracts current interest, it is yet to be used in spectrum management, even though transparent and regulatively acceptable decision-making is required. The security issues like adversarial attacks and data poisoning have the potential of reducing the performance of the system and require secure and robust frameworks. FL and blockchain technologies have been suggested to facilitate decentralized and privacy-preserving spectrum sharing (Zuo et al., 2023), although there are vulnerabilities. Also, new heterogeneous and high-frequency networks (e.g., mmWave, THz, RIS-assisted, and space-air-ground systems) also present such complexities as the rapid variations of channels and strict latency requirements (Wei et al., 2024). The future work should be on the hybrid models that can balance interpretability and performance, lightweight and compressible AI methods that can be deployed on the real-time, and secure FL framework that can counter the adversarial threats and assure scalability and performance in the next generation of wireless networks.

CONCLUSION

The wireless gadgets and the new applications like IoT, V2X communication, and immersive VR/AR services have exerted an unprecedented strain on the radio spectrum. The modern wireless systems are too dynamic, uncertain, and heterogeneous to be dealt with using traditional statistical and optimization methods. AI and ML have thus become effective spectrum prediction and optimization tools. Specifically, DL models like CNNs and RNNs are better than traditional ones in their ability to capture complex patterns in the temporal and spatial spectrums whereas RL/DRL allows allocating resources adaptively and model-free.

Prediction and optimization integration is a major step forward and networks are capable of dynamically determining the availability of spectrum, and make real-time allocation decisions. FL and multi-agent systems techniques ensure a high level of scalability, privacy, and flexibility, whereas blockchain and adversarially robust AI can help in enhancing the security of agents in uncertain environments (Ning et al., 2024). In the future, these strategies will be required in 6G networks with ultra-dense connectivity and multiple applications. Nonetheless, issues like computational problems, privacy, adversarial resistance, and model elucidation must be countered to allow dependable and intelligent spectrum control.

PRACTITIONERS TAKEAWAY

Practically, AI/ML-based spectrum prediction and optimization allow to make the allocation of spectrum in next-generation wireless systems more efficient and intelligent. Prediction is based on data which makes it possible to make proactive choices on channel availability and interference which is better than reactive schemes. The quality of these techniques and methods relies on good data and preprocessing. The complex spectrum patterns well represented by DL models include CNNs, RNNs, LSTM, and transformers but should be chosen relative to computational and latency costs, particularly in an edge and Internet of Things deployment. Blended models based on both conventional optimization and ML provide stable and comprehensible answers. RL can be used to aid dynamic allocation, whereas FL can be used to improve scalability and privacy. Prediction and optimization are useful in enhancing spectral efficiency and QoS. Nonetheless, some issues such as the complexity of calculations, data needs, interpretability, and security should be tackled to be deployed in practice.

REFERENCES

Akyildiz,I. F., W. Y. Lee, M. C. Vuran, and S. Mohanty, "Next generation/dynamic spectrum access/cognitive radio wireless networks: A survey," *Computer Networks*, vol. 50, no. 13, pp. 2127–2159, Sep. 2006.

Bi, S., L. Huang, and Y.-J. A. Zhang, "Joint optimization of spectrum sharing and power allocation in cognitive radio networks via deep reinforcement learning," *IEEE Transactions on Wireless Communications*, vol. 20, no. 2, pp. 1182–1194, Feb. 2021.

Chen, D., Y. Liu, and J. Li, "Multi-user dynamic spectrum access based on distributed dueling double deep Q-network," in *Proc. ACM/IEEE*, Oct. 2024.

Chen, M., Z. Yang, W. Saad, C. Yin, H. V. Poor, and S. Cui, "A joint learning and communications framework for federated learning over wireless networks," *IEEE Transactions on Wireless Communications*, vol. 20, no. 1, pp. 269–283, Jan. 2021.

Chen, R., Z. Lu, Z. Ding, and Y. Qian, "Deep learning for spectrum prediction in cognitive radios," *IEEE Wireless Communications Letters*, vol. 9, no. 11, pp. 1894–1898, Nov. 2020.

Dang, S., O. Amin, B. Shihada, and M.-S. Alouini, "What should 6G be?," *Nature Electronics*, vol. 3, no. 1, pp. 20–29, Jan. 2020.

Deshpande, R. D., F. A. Khan, and Q. Z. Ahmed, "Spectrum sharing using deep reinforcement learning in vehicular networks," arXiv preprint arXiv:2410.12521, 2024.

Elsayed, M. and M. Erol-Kantarci, "AI-enabled future wireless networks: Challenges, opportunities, and open issues," *IEEE Vehicular Technology Magazine*, vol. 14, no. 3, pp. 70–77, Sep. 2019.

Elsayed, M., A. Thakur, and M. Erol-Kantarci, "Reinforcement learning for advanced wireless networks: A survey," *IEEE Communications Surveys & Tutorials*, vol. 23, no. 4, pp. 2790–2823, 4th Quart., 2021.

He, Y., N. Zhao, and H. Yin, "Integrated resource allocation for 6G: Potential, principles, and prospects," *IEEE Wireless Communications*, vol. 29, no. 4, pp. 98–105, Aug. 2022.

Jagannath, A., J. Jagannath, and T. Melodia, "Redefining wireless communication for 6G: Signal processing meets deep learning with deep unfolding," *IEEE Transactions on Artificial Intelligence*, vol. 2, no. 6, pp. 528–536, Dec. 2021.

Letaief, K. B., W. Chen, Y. Shi, J. Zhang, and Y.-J. A. Zhang, "The roadmap to 6G: AI empowered wireless networks," *IEEE Communications Magazine*, vol. 57, no. 8, pp. 84–90, Aug. 2019.

Li, L., Y. Zhao, and X. Wang, "A hierarchical federated deep reinforcement learning approach for dynamic spectrum access in IoT," *IEEE Internet of Things Journal*, early access, 2024.

Lu, Y., X. Huang, K. B. Letaief, and J. Zhang, "Blockchain and federated learning for secure spectrum sharing in 6G," *IEEE Network*, vol. 35, no. 4, pp. 128–135, Jul. 2021.

Mitola, J. and G. Q. Maguire, "Cognitive radio: Making software radios more personal," *IEEE Personal Communications*, vol. 6, no. 4, pp. 13–18, Aug. 1999.

Mumtaz, S., J. Rodriguez, and L. Dai, *MmWave Massive MIMO: A Paradigm for 5G*, London, U.K.: Academic Press, 2016.

Ning, W., J. Li, and X. Tang, "Blockchain-based federated learning: A survey and new perspectives," *Applied Sciences*, vol. 14, no. 20, 9459, 2024.

O'Shea, T. and J. Hoydis, "An introduction to deep learning for the physical layer," *IEEE Transactions on Cognitive Communications and Networking*, vol. 3, no. 4, pp. 563–575, Dec. 2017.

Pan, G., Q. Wu, B. Zhou, J. Li, W. Wang, G. Ding, and D. K. Y. Yau, "Spectrum prediction with deep 3D pyramid vision transformer learning," arXiv preprint arXiv:2408.10120, 2024.

Qamar, F., S. A. Khan, and A. A. Khan, "Federated learning for millimeter-wave spectrum in 6G networks: Applications, challenges, and open issues," *PeerJ Computer Science*, vol. 10, 2024.

Qin, Z., H. Ye, G. Y. Li, and B.-H. Juang, "Deep learning in physical layer communications," *IEEE Wireless Communications*, vol. 26, no. 2, pp. 93–99, Apr. 2019.

Sabir, H., A. M. Rahman, and P. Kumar, "A systematic review of AI-enabled spectrum management in 6G: Datasets, benchmarks, and open challenges," arXiv preprint arXiv:2407.10981, 2024.

Sagduyu, S., Y. Shi, and T. Erpek, "Adversarial machine learning for wireless communications and networking," *IEEE Communications Magazine*, vol. 57, no. 11, pp. 115–121, Nov. 2019.

Sun, H., A. Nallanathan, C.-X. Wang, and Y. Chen, "Wideband spectrum sensing for cognitive radio networks: A survey," *IEEE Wireless Communications*, vol. 20, no. 2, pp. 74–81, Apr. 2013.

Ukpong, U. C., A. K. Muhammad, and M. R. Akbar, "Deep reinforcement learning agents for dynamic spectrum access in television whitespace cognitive radio networks," *Scientific African*, vol. 27, e02523, pp. 1–16, early access, 2025.

Wang, X., L. Gao, and S. Mao, "Phase retrieval for spectrum sensing in cognitive radio networks: A deep learning approach," *IEEE Transactions on Cognitive Communications and Networking*, vol. 6, no. 1, pp. 43–55, Mar. 2020.

Wen, X., Z. Ning, J. Han, Y. Wang, and H. Li, "Joint multidimensional pattern for spectrum prediction using GNN," *Sensors*, vol. 23, no. 21, Art. no. 8883, Nov. 2023.

Xu, Y., J. Wang, Q. Wu, A. Anpalagan, and Y. D. Yao, "Opportunistic spectrum access in cognitive radio networks: Global optimization using local interaction games," *IEEE Journal on Selected Topics in Signal Processing*, vol. 6, no. 2, pp. 180–194, Apr. 2012.

Ye, H. and G. Y. Li, "Deep reinforcement learning for resource allocation in V2V communications," *Proc. IEEE ICC*, pp. 1–6, 2018.

Ye, H., G. Y. Li, and B.-H. Juang, "Deep reinforcement learning based resource allocation for V2V communications," *IEEE Transactions on Vehicular Technology*, vol. 68, no. 4, pp. 3163–3173, Apr. 2019.

Zappone, A., M. Di Renzo, and M. Debbah, "Wireless networks design in the era of deep learning: Model-based, AI-based, or both?," *IEEE Transactions on Communications*, vol. 67, no. 10, pp. 7331–7376, Oct. 2019.

Zhang, C., P. Patras, and H. Haddadi, "Deep learning in mobile and wireless networking: A survey," *IEEE Communications Surveys & Tutorials*, vol. 21, no. 3, pp. 2224–2287, 3rd Quart., 2019.

Zhang, S., H. Dai, and S. K. Das, "Dynamic spectrum access for Internet-of-Things with hierarchical personalized federated deep reinforcement learning," *Ad Hoc Networks*, vol. 160, Art. no. 103257, Nov. 2023.

Zhao, N., Y. Liang, and D. Niyato, "Deep reinforcement learning for dynamic spectrum access: A review," *IEEE Communications Magazine*, vol. 59, no. 1, pp. 46–52, Jan. 2021.

Zhu, H., X. Li, and Y. Chen, "ADP-GNN: A spectral graph neural network framework for high-resolution spectral learning tasks," *Journal of Systems Architecture*, 168, early access, 2025.

Zorzi, M., A. Zanella, A. Testolin, M. D. Penrose, and M. Zorzi, "Cognition-based networks: A new perspective on network optimization using learning and distributed intelligence," *IEEE Network*, vol. 31, no. 5, pp. 44–51, Sep. 2017.

CHAPTER 6

Machine Learning-Based Spectrum Forecasting for Ultra-Dense IoT Environments

Jyoti Sunil More, Bharati Khatawate, Nupur Gaikwad, and Deepshikha Sarjare

6.1 INTRODUCTION

6.1.1 Background

With the advent of digital transformation in multiple domains which needs integration of devices, sensors, etc., with applications like healthcare, industrial automation, smart homes, and transportation, there is need of major transformations in Internet of Things (IoT). [3][6]

The inefficient utilization leads to restrictions on the performance of modern wireless systems, when the available frequency bands ae limited. Conventional static allocation approaches where typical frequencies are assigned to specific services leads to challenges regarding the struggle to adapt to evolving and uncertain communication patterns.

Due to the overlapping transmissions, issues such as interference, delays, congestion thereby causing degraded QoS may result. IoT systems demand low latency and high throughput and reliability for dense networks.

DOI: 10.1201/9781003731597-6

6.1.2 Spectrum Forecasting: Concept and Importance

The prediction of how radio frequency channels will be used over the forthcoming time intervals is done in the spectrum forecasting.

The conventional sensing methods, usually respond only to the concerned environment which is critical in ultra dense IoT settings. Collisions can be reduced by expecting idle periods, by optimizing scheduling and minimizing unwanted energy consumption. The emerging technologies like cognitive radio (CR) and the artificial IoT support the spectrum forecasting to a great extent.

6.1.3 Role of Machine Learning (ML) in Spectrum Forecasting

To aid the unpredictable, multidimensional, and nonlinear nature of wireless spectrum data, various customized tools are provided by machine learning (ML) [1][4]. ML techniques directly get trained from real world data, supported by interference, mobility, and environmental changes, unlike classical statistical models that assume stable and predictable behavior. Reinforcement learning (RL) and deep learning (DL) to predict channel availability, locate inactive bands, and check the interference risks more accurately along with revealing essential and complex temporal and spatial dependencies.

6.1.4 Motivation for Ultra-Dense IoT Environments

The dense IoT networks are identified differently as non-conventional wireless systems, due to the scale of network, diversity and bursty traffic patterns. Dynamic behavior becomes challenging for the traditional spectrum allocation mechanisms, typically the ones relying on static allocations and centralized control. This leads to wastage of spectrum resources and increase in interference, since they respond very slowly to the rapid fluctuations [7]. In such cases, spectrum forecasting based on ML is an effective alternative.

6.2 SPECTRUM UTILIZATION AND FORECASTING IN IOT NETWORKS

6.2.1 Overview of IoT Network Architecture

IoT is interpreted as a broad, interconnected platform that integrates edge and cloud platforms, smart devices and communication gateways together. Many IoT communication models [16]–[20] showcase framework as combination of three fundamental layers, as shown in Figure 6.1.

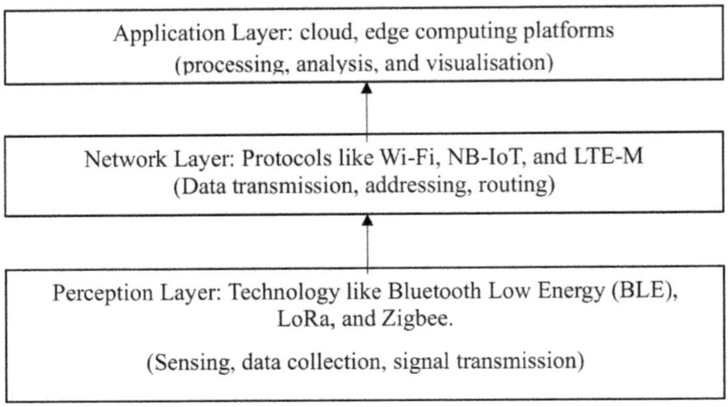

FIGURE 6.1 IoT communication framework.

Perception layer is the fundamental layer is involved in sensing and acquiring data. The network layer deals with device addressing, routing, and data transmission. Application layer is the topmost layer, providing services to end users by utilizing cloud and edge computing platforms. The network layer gets a significant load, with increase in number of IoT devices involved. The congestion becomes unavoidable, due to enormous involvement of devices attempting to communicate at a time.

6.2.2 Spectrum Utilization Patterns in Dense IoT Scenarios

Time and location widely affect the usage of spectrum in IoT networks. IoT communication has tendency to be irregular. Whenever there is any trigger like sensor detecting movement, a user interacts with device or if any change occurs in the environment, the IoT devices sends data in that event. Thus, due to this bursty transmission, some parts of spectrum become congested while other channels remain idle for long periods.

6.2.3 Spectrum Forecasting: Definition and Conceptual Foundation

The process of analyzing the past usage trends and environmental factors and predicting if the given radio channel will be available or occupied for the given time in future is known as spectrum forecasting. A general structure of a spectrum forecasting system is depicted in Figure 6.2. The spectrum usage exhibits known temporal and spatial patterns, in spite of the fact that it seems to be random. The goal of spectrum forecasting is to predict the future state x_t+k_x of the channel that signifies if the channel is occupied or free after the time interval k, based on current state x_t. The main objectives of spectrum forecasting include improve spectrum efficiency (η), reduce collision probability (P_e), reduce latency (L), and enhance Quality of Service (QoS)

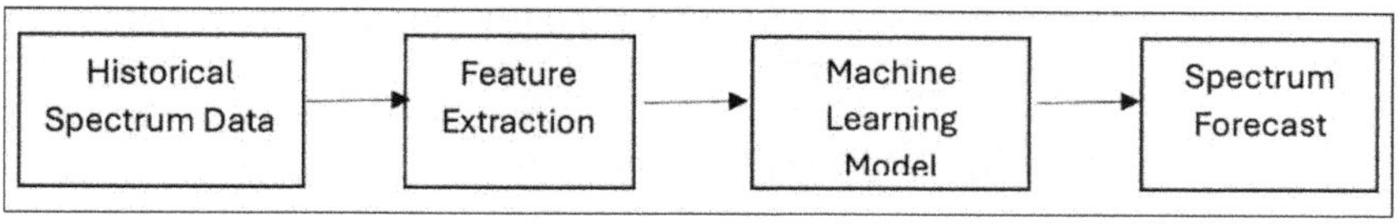

FIGURE 6.2 Block diagram representing spectrum forecasting model.

TABLE 6.1 Comparison of Traditional and ML-Based Spectrum Forecasting 1 Approaches

Approach	Methodology	Adaptability	Computational Complexity	Scalability in IoT	Forecast Accuracy
Static	Fixed band assignment	Low	Low	Poor	Low
HMM/ ARIMA	Statistical prediction	Moderate	Moderate	Limited	Moderate
Cognitive DSA	Rule-based access	Moderate	High	Limited	Moderate
ML	Data-driven pattern	High	Moderate	Good	High
Deep Learning	Temporal-spatial	Very High	High	Very Good	Very High
RL	Adaptive control	Very High	High	Excellent	High
FL	Distributed collaborative	High	Moderate	Excellent	High

6.2.4 Traditional Spectrum Management Approaches

The spectrum management traditional methods highly relied on traditional, rule-based approaches, based on signal processing, typically before the integration of AI [1][14][21][22] and ML. These typical strategies are classified into the classes, viz.., static spectrum allocation (SSA), dynamic spectrum access (DSA), autoregressive and Markovian models, and optimization and game-theoretic approaches.

6.2.5 Transition Toward ML-Based Spectrum Prediction

ML has enabled to get rid of rigidness and give the model-based strategies towards flexible, data driven ones [2]. Nonlinear learning, adaptability, generalization, and automation are some advantages of ML-based spectrum predictions.

As compared to conventional methods, the modern technologies like Reinforcement Learning (RL), Recurrent Neural Networks (RNNs), and Long Short-Term Memory (LSTM) networks [23]–[25] outperform efficiently, especially for tasks containing sequential or runtime input. The comparison of traditional and ML approaches is shown in table 6.1.

6.3 ML FOUNDATIONS FOR SPECTRUM FORECASTING

6.3.1 Overview of ML Techniques Applicable to Spectrum Prediction

ML has significantly modified the interpretation and prediction of spectrum behavior in Internet of Things (IoT) environments. By referring to the past channel activity, signal characteristics, and contextual data, such as time of day, device density, or geographic location, the ML models may be able to give more accurate predictions of future spectrum conditions. The four major learning paradigms for ML based spectrum techniques are supervised learning, unsupervised learning, reinforcement learning, and deep learning [4][26].

6.3.2 Supervised Learning Models

These ML methods are compared in Table 6.2.

6.3.3 Unsupervised and Semi-Supervised Models

Gathering well-labeled spectrum datasets is time-consuming and resource-intensive in real-world IoT implementations. Semi-supervised methods [5] combine a limited amount of modelling data with huge pools of unlabeled samples to enhance predictive accuracy, unsupervised models reveal underlying structure in unlabeled data.

Clustering Techniques:

Based on similarities in their temporal or frequency-domain behavior, clustering methods like K-means, Gaussian Mixture Models (GMM), and DBSCAN classify spectrum observations.

TABLE 6.2 Comparison of ML Models

Model	Type	Advantages	Limitations	Use Case
Linear Regression	Regression	Simple, interpretable	Fails for nonlinear data	Predict signal power
SVM	Classification	Robust, effective in high-dimensional space	Sensitive to kernel choice	Binary occupancy prediction
Random Forest	Classification/ Regression	High accuracy, low overfitting	Requires more computation	Dynamic spectrum access
kNN	Classification	Simple, instance-based	Poor for large-scale data	Small network predictions

Dimensionality Reduction:
The high-dimensional data is reduced into more condensed and significant representations using methods like Principal Component Analysis (PCA) and t-distributed Stochastic Neighbor Embedding (t-SNE).

Semi-Supervised Learning:
When modelling data is scarce, semi-supervised methods become especially useful. Algorithms such as Label Propagation and Co-Training draw on relationships among unlabeled samples and the limited known labels to infer missing channel states.

6.3.4 Reinforcement Learning for Dynamic Spectrum Access (DSA)

Each IoT node acts as an agent in an RL context. It learns to select the optimal channel through interactions with the wireless environment and also the feedback in the form of rewards or penalties. These include Q-learning, Deep Q-Networks(DQN), Multi Agent Reinforcement Learning (MARL).

6.3.5 Deep Learning and Temporal Models

The capabilities of spectrum prediction have been greatly increased by Deep Learning (DL) approaches, particularly in situations where the wireless environment displays intricate spatial and temporal dynamics.

Recurrent Neural Networks (RNNs):
RNNs retain a type of internal memory that transfers data from one time step to the next in order to handle consecutive inputs. Because of this characteristic, they can be used to mimic periodic variations in spectrum consumption or recurrent patterns.

Long Short-Term Memory (LSTM) and Gated Recurrent Units (GRU):
Through gating mechanisms that control how information is kept and forgotten, LSTM and GRU models solve many of the drawbacks of conventional RNNs.

Convolutional Neural Networks (CNNs):
CNNs have shown promise in spectrum forecasting when spectrograms or frequency-domain representations are utilized as input, despite their widespread association with picture or spatial pattern recognition.

TABLE 6.3 Performance Metrics and the Interpretations

Metric	Formula/Definition	Interpretation
Accuracy (A)	A = {TP + TN}/{TP + TN + FP + FN}	Proportion of correct predictions
Precision (P)	P = {TP}/{TP + FP}	Reliability of predicted busy/idle states
Recall (R)	R = {TP}/{TP + FN}	Ability to detect true channel states
F1 Score	$F1 = 2 \times \frac{P \times R}{P+R}$	Balance between precision and recall
RMSE	$RMSE = \sqrt{\frac{1}{N}\sum (y_t - \hat{y}_t)^2}$	Prediction error for regression tasks
Prediction Horizon (H)	Time steps forecasted ahead	Measures model's temporal foresight

Hybrid Deep Learning Models:

Combining CNNs with LSTMs to take use of their respective strengths—CNNs are good at extracting spatial characteristics, while LSTMs are better at modelling temporal sequences—is becoming more and more popular.

6.3.6 Performance Metrics for Spectrum Forecasting

The effectiveness of ML-driven spectrum prediction models is typically assessed using a range of quantitative metrics, which are organized in Table 6.3.

6.4 SUPERVISED LEARNING TECHNIQUES FOR SPECTRUM PREDICTION

6.4.1 Overview

Spectrum Prediction is a technique that is used to forecast the spectrum's next state based on current and historical observations. Once trained, these models can provide both short-term and long-term predictions that support efficient DSA.

6.4.2 Regression-Based Models

These models use statistical relationship to forecast spectral properties like power and occupancy or related physical attributes like weight from spectral data by mapping input such as time, physical parameters to spectral output.

6.4.2.1 *Linear Regression*

Linear regression models the target value as a weighted sum of input features: $y = \beta_0 + \sum_{i=1}^{n} \beta_i x_i + \varepsilon$ where βi are model coefficients and ϵ is the error term.

6.4.2.2 *Support Vector Regression (SVR)*

SVR performs effectively on small to medium-sized datasets and in environments with nonlinear or noisy measurements. Its flexibility makes it suitable for moderately dense, non-stationary IoT systems.

6.4.2.3 *Random Forest Regression*

For heterogeneous IoT datasets with both continuous and categorical inputs, like radio bands, time slots, and device types, RF models work incredibly well.

6.4.2.4 *Gradient Boosting Regression*

They are appropriate for dynamic and interference-prone IoT contexts because they can simulate intricate nonlinear relationships among spectrum characteristics.

6.4.3 Temporal Prediction Using Time-Series Models

Predicting the temporal network such as a physical contact network in the future is essential to forecast performance of a process upon the network spectrum occupancy often exhibits temporal dependencies—for instance, certain frequency bands are more active during specific times of day due to repeating IoT communication patterns.

6.4.3.1 *ARIMA (Autoregressive Integrated Moving Average)*

ARIMA is most effective when spectrum usage patterns are stable or periodic, but its performance declines in nonlinear, bursty, or highly dynamic IoT environments. It often serves as a baseline for evaluating more advanced deep learning methods.

6.4.3.2 *Long Short-Term Memory (LSTM) Networks*

Empirical studies reveal that LSTM-based predictors outperform ARIMA and SVR in terms of Root Mean Square Error (RMSE), particularly in modelling energy levels and idle channel probabilities.

6.4.3.3 Gated Recurrent Units (GRUs)

By combining input and forget gates into a single update gate, GRUs reduce computational complexity and simplify the LSTM architecture.

6.4.4 CNN-Based Models for Learning Spatial–Frequency Patterns

Convolutional Neural Networks (CNNs) are particularly good at identifying spatial-frequency correlations in spectrum data. CNNs may identify localized features like interference bursts, spectrum holes, and recurring occupancy trends when spectrum measurements are displayed as spectrograms, which are plots of frequency vs time and power intensity.

6.5 UNSUPERVISED AND SEMI-SUPERVISED LEARNING IN SPECTRUM ANALYSIS

6.5.1 Introduction

The performance of supervised learning models have a significant dependency on availability of labeled and large dataset. These approaches have capability to gather, analyze and interpret spectrum data with minimum manual labelling.

6.5.2 Clustering for Spectrum Usage Pattern Identification

It is one of the most basic unsupervised learning approaches is clustering, wherein it groups data points based on similar characteristics.

6.5.2.1 K-Means Clustering

The variance within each cluster is minimized by using K-Means, a popular clustering algorithm that partitions data into k clusters: $J = \sum_{i=1}^{k} \sum_{x_j \, \varepsilon \, C_i}^{k} \|x_j - \mu_i\|^2$ where, μ_i represents the centroid of cluster $C_{i.}$ Each data vector xj denotes features like frequency band, received signal strength (RSS), and occupancy duration, which are significant in spectrum analysis.

6.5.2.2 Density-Based Clustering (DBSCAN)

The limitation of centroid based clustering is that it identifies groups based on distance. To overcome this, there is another approach i.e. DBSCAN where, grouping is based on data density. DBSCAN efficiently separates heavily used and idle spectrum regions, allowing customized channel allocation.

6.5.2.3 *Hierarchical and Model-Based Clustering*

The spectrum similarities at different hierarchical levels can be shown using a tree like structure called dendrogram and used for interpreting the relationships between frequency bands and devices. It deploys the Gaussian Mixture Models (GMM) for model based clustering.

6.5.3 Autoencoders for Anomaly Detection in Spectrum Occupancy

The autoencoders can detect anomalies such as unexpected energy spikes, jamming attempts or illegal transmissions in spectrum analysis. Autoencoders are extensively used in CR networks specially for anomaly and intrusion detection.

6.5.4 Semi-Supervised Learning for Limited Labeled Data

The strengths of supervised and unsupervised methods can be combined with semi-supervised learning by using both labeled and unlabeled data during training.

6.5.4.1 *Self-Training and Pseudo-Labelling*

Self-training means a model is first trained on a small labeled dataset, eventually it predicts labels for unlabeled samples. These are added back into training set and are referred as pseudo-labels.

6.5.4.2 *Graph-Based Semi-Supervised Learning*

This approach is specifically effective in the environments where the spatially correlated IoT networks are designed. Here the nearby devices exhibit similar transmission characteristics.

6.5.4.3 *Semi-Supervised Deep Learning*

Recent frameworks based on semi supervised learning gather deep neural networks with consistency regularization, to ensure that small input customizations help to get stable predictions. This improves scalability and reduces dependency on wide manual labelling. This is advantageous in dynamic IoT environments. The comparison is shown in table 6.4.

6.5.5 Dimensionality Reduction for Spectrum Visualization

The spectrum data with high dimensionality having multiple frequency bands, time intervals and spatial readings can lead to difficulty in interpreting.

TABLE 6.4 Comparative Analysis of Model Performance in Dynamic IoT Settings

Model	Type	Strengths	Limitations	Best Application Scenario
Linear Regression	Regression	Simple, interpretable, low computation	Poor nonlinear handling	Low-density, stationary IoT environments
SVR	Regression	Good for small nonlinear datasets	High training cost for large data	Small-scale or local spectrum forecasting
Random Forest	Ensemble Regression	Robust, handles feature diversity	Slower inference with many trees	Multi-band or heterogeneous
Gradient & XGBoost	Ensemble Regression	High accuracy, complex relationships	Sensitive to tuning, large memory use	Dense and dynamic IoT networks
ARIMA	Time-Series	Simple, interpretable	Ineffective for non-stationary data	Baseline forecasting model
LSTM	Deep Learning	Captures long-term dependencies	Training-intensive	Temporal spectrum forecasting
GRU	Deep Learning	Efficient and lightweight	lower accuracy than LSTM	Edge/IoT node prediction
CNN	Deep Learning	Captures spatial-frequency features	Requires large data, GPU	Multi-channel

6.5.5.1 Principal Component Analysis (PCA)

Important for feature compression before regression based prediction or clustering.

6.5.5.2 t-Distributed Stochastic Neighbor Embedding (t-SNE)

Highlights groups of channels or devices having more similarity index, which helps it to identify related frequency bands.

6.6 REINFORCEMENT LEARNING FOR DYNAMIC SPECTRUM DECISION-MAKING

6.6.1 Introduction

Through interaction, agents (such base stations or Internet of Things nodes) acquire appropriate spectrum access behaviors via Reinforcement

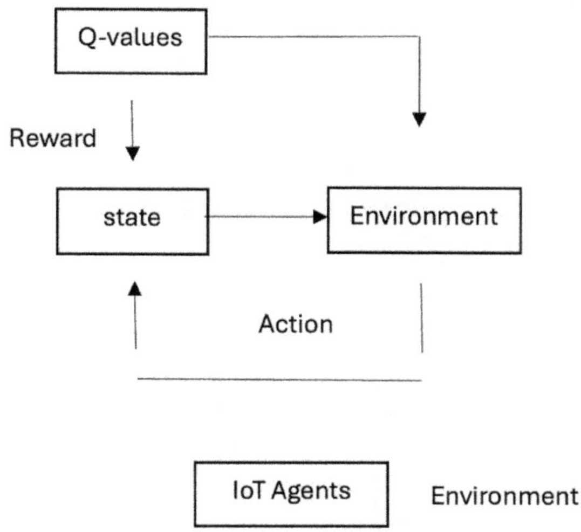

FIGURE 6.3 Basic Reinforcement learning framework for spectrum forecasting.

Learning (RL) [8]. The fundamental RL structure for spectrum forecasting is shown in Figure 6.3.

Unlike supervised approaches, reinforcement learning is independent of pre-labeled datasets. Reinforcement learning provides autonomous, adaptable spectrum management that improves consumption, lowers collision rates, and tailors decisions depending on current conditions in quickly evolving IoT contexts.

6.6.2 Fundamentals of Reinforcement Learning in Spectrum Management

RL problems are commonly framed as Markov Decision Processes (MDPs). An MDP for spectrum management typically includes:

- State (S): A representation of the radio environment
- Action (A): The decision taken by the agent
- Reward (R): Scalar feedback measuring success
- Policy (π): The mapping from states to actions that the agent seeks to optimize.
- Transition probability (P): The environmental factors

The agent's goal is to learn an optimal policy π^* that maximizes expected discounted return:

$$Q*(s,a) = \max_{\pi} \mathrm{E}\left[\sum_{t=0}^{\infty} \gamma^{t} r_{t+1} \,|\, s_t = s, a_t = a, \pi\right]$$

where $0<\gamma<1$ is the discount factor. An IoT spectrum forecasting agent constantly observes the radio scene, forecasts availability, and adapts transmission choices accordingly.

6.6.3 Q-Learning for Spectrum Access Optimization

Q-Learning is a model-free algorithm that learns the action-value function Q(s,a) in the absence of a transition model for the environment. Q-values are updated iteratively via the Bellman rule.

6.6.4 Deep Q-Networks (DQN) for Spectrum Forecasting

DQN [10] extends Q-Learning by approximating the Q-function with a neural network

6.6.4.1 DQN Architecture for Spectrum Analysis

A typical DQN for spectrum tasks includes:

- Input layer: Encodes features like recent occupancy history, interference metrics, etc.
- Hidden layers: Convolutional or recurrent units extract spatio-temporal patterns.
- Output layer: Produces Q-values for each possible action (e.g., candidate channels).

6.6.4.2 Experience Replay and Target Networks

DQN stabilizes training using two key techniques:

- Experience replay: Store past transitions (s, a, r, s′) in a buffer, then train on randomized mini-batches in order to break correlation between samples.
- Target network: Maintain a periodically updated copy of the Q-network to reduce oscillations and divergence during learning.

6.6.4.3 *Use in Dense IoT Applications*

From robust channel selection policies to high-dimensional spectrum observations, the Deep Q-Network can learn intricate mappings. In nonstationary conditions, DQN usually performs better than classical algorithms, although it needs more computation and careful tweaking.

6.6.5 Multi-Agent Reinforcement Learning (MARL) for Distributed Spectrum Sharing

Multi-Agent Reinforcement Learning (MARL) [12][13] provides a decentralized approach because centralized control becomes unfeasible at scale.

6.6.5.1 *Cooperative vs. Competitive MARL*

In Cooperative MARL, the agents share information or rewards whereas in Competitive MARL the agents act solely.

6.6.5.2 *Communication and Coordination*

Parameter sharing and Centralized Training with Decentralized Execution (CTDE), in which agents are trained with access to global information but operate independently during runtime, are common MARL techniques.

6.6.5.3 *MARL Algorithms for Spectrum Access*

Well-known approaches include: Independent Q-Learning (IQL), MADDPG (Multi-Agent DDPG) and QMIX. In massive IoT deployments where centralized control is impractical, these techniques enable distributed decision-making.

6.6.6 Reward Function Design for IoT Spectrum Optimization

In addition to encoding the trade-offs between throughput, energy consumption, fairness, and interference, the reward function guides agent behavior. Stable learning and practical policies require careful incentive design, particularly in multi-agent environments where feedback may be noisy or delayed.

6.6.6.1 *Typical Reward Components*

These include throughput reward, encourages choosing high-quality channels, energy efficiency term, Fairness metric, etc.

6.6.6.2 Composite Reward Design

A practical reward often combines components in a weighted sum with weights chosen to reflect network priorities (e.g., throughput vs. interference control).

6.7 FEDERATED AND EDGE-ASSISTED LEARNING FOR DISTRIBUTED SPECTRUM INTELLIGENCE

6.7.1 Introduction

Traditional centralized methods of spectrum forecasting confront significant challenges when IoT networks grow into ultra-dense environments. Federated learning (FL) [9] [15][27]–[29] and Edge-Assisted Learning [30] have evolved as creative solutions to these problems.

6.7.2 Federated Learning (FL) for Spectrum Forecasting

A distributed ML framework that permits multiple clients like IoT devices, base stations, or gateways to collectively train a generic model without sharing raw data is termed as FL. The central server later aggregates them into a refined global model as depicted in Figure 6.4

6.7.2.1 Basic FL Framework

1. Initialization: The central server initializes a global model W_0
2. Local Training: The model is updated by each client k, using its local spectrum data for multiple epochs: $W_k^{t+1} = W_k^t - n \nabla L_k (W_k^t)$

where L_k is the local loss function and η is the learning rate.

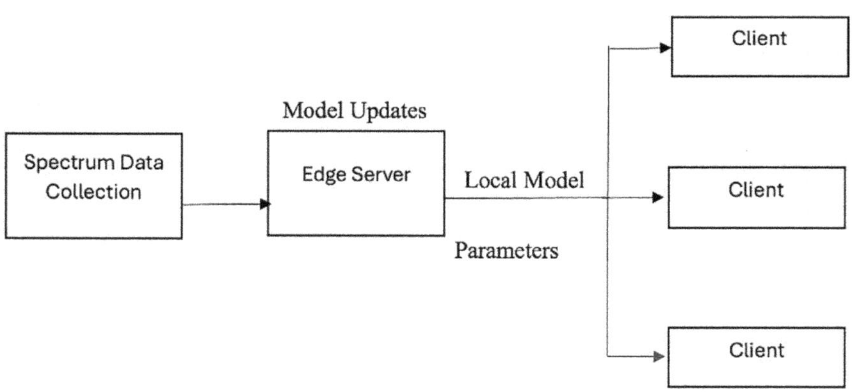

FIGURE 6.4 Federated edge assisted learning architecture for spectrum forecasting.

3. Model Aggregation: The server performs Federated Averaging (FedAvg) to combine client updates: $W^{t+1} = \sum_{k=1}^{k} \frac{n_k}{n} W_k^{t+1}$

where n_k represents the number of samples at client k, and $n=\sum_k n_k$ is the total data size.

4. Iteration: The global model is redistributed for further local training until convergence.

6.7.2.2 *Applications in Spectrum Forecasting*

- Distributed Channel Prediction: The spectrum occupancy and interference of signals is predicted by training local models by IoT devices.
- DSA: IoT devices can forecast idle channels with the help of federated models and help to avoid collisions in real time, by supporting efficient spectrum allocation

6.7.3 Edge-Assisted Learning for Low-Latency Spectrum Intelligence

Edge-assisted learning builds upon federated concepts by offloading computation and model aggregation to edge nodes (e.g., gateways or access points). This localized processing minimizes latency and dependence on remote cloud servers.

6.7.3.1 *Edge-Based Architecture:*

It includes prominent stages like data collection, edge processing, model aggregation and decision deployment. For use cases like industrial IoT, autonomous driving, and smart sensor systems the Real-time forecasting, is essential and is supported by this architecture. The key advantages are low latency, real-time decision-making, and energy efficiency.

6.7.3.2 *Example Use Case:*

In a smart city, neighboring edge servers gather spectrum data from street sensors and traffic cameras. These servers offer low-latency, collision-free spectrum access without need of continuous cloud connectivity by predicting the available channels for every device cluster.

6.7.4 Hybrid Federated–Edge Learning

To achieve hierarchical spectral intelligence, a hybrid learning architecture combines both federated and edge-assisted paradigms. This layered system combines privacy preservation, distributed computation, and low-latency response. This architecture provides an enhanced scalability for large-scale IoT ecosystems, lowered communication overhead between devices and the cloud, and faster adaptability to localized spectrum modifications.

6.7.5 Privacy-Preserving Techniques

Safeguarding privacy is a core requirement in federated and edge-assisted systems, as spectrum usage data can indirectly reveal sensitive user or device behavior. Common privacy-enhancing methods include: differential privacy (DP), secure aggregation, and homomorphic encryption. Together, these techniques protect user and device privacy while enabling collaborative model training across large IoT networks.

6.7.6 Performance Evaluation Metrics

The efficiency and effectiveness of federated and edge-assisted spectrum learning are assessed using key performance metrics such as prediction accuracy, spectrum utilization efficiency (SUE), convergence time, latency, and energy consumption.

6.8 SIMULATION FRAMEWORKS AND PERFORMANCE EVALUATION FOR IOT SPECTRUM FORECASTING

6.8.1 Introduction

In ultra-dense IoT setup, for developing a trusted machine learning model for spectrum prediction, it is very important to do rigorous testing to ensure the accuracy, scalability and robustness. It is highly impractical to conduct real world experiments involving multiple IoT devices. This is mainly because of high costs, technical complexity, and strict management needs. To address these challenges, the simulation architectures are commonly used as alternatives.

6.8.1.1 Standards and Regulatory Considerations in ML-Based Spectrum Management

Spectrum forecasting using machine learning must follow internationally approved international technical and regulatory norms. IEEE 802.22, describe CR-based WRANs that function in TV white spaces and offer database assisted coordination and DSA techniques. Some of the similar improvements are made to NR-U (New Radio in Unlicensed Spectrum) in 3GPP Release 17, which is allowed for intelligent coexistence in shared bands.

European efforts to harmonize spectrum directs by European Telecommunications Standards Institute (ETSI) through frameworks for spectrum sharing and Licensed Shared Access (LSA). In the United States, the Federal Communications Commission (FCC) oversees Citizens Broadband Radio Service (CBRS) which is used to coordinate the sharing in dynamic access systems in Spectrum access systems (SAS).

6.8.2 Simulation Frameworks for Spectrum Forecasting

To showcase the real world IoT network in the simulated environment, usually the simulation architectures provide a configurable, customized environment.

6.8.2.1 NS-3 (Network Simulator 3)

It offers the services like in depth modelling of physical and MAC layer protocols, modules for CR simulations, supporting DSA and capabilities for generating synthetic datasets that capture spectrum occupancy and interference trends.

6.8.2.2 MATLAB/Simulink

MATLAB provides a high-level support for algorithm design, analysis, and visualization. It supports direct implementation of models such as ARIMA, LSTM, CNN, and reinforcement learning frameworks for spectrum predicting tasks.

6.8.2.3 SUMO (Simulation of Urban Mobility)

SUMO serves as a significant tool for IoT systems that involve mobility, such as connected vehicles or smart transportation networks. SUMO aids in evaluating forecasting algorithms under dynamic, spatially distributed communication conditions, when combined with NS-3 or OMNeT++.

6.8.2.4 Cognitive Radio (CR) Testbeds and Emulators

To test machine learning models under wireless conditions, the physical testbeds are provided using software defined radios (SDRs), such as

USRP-based systems. They produce valuable real-world data on interference, signal propagation and hardware level performance but exhibit more resource demanding than simulations.

6.8.3 Key Evaluation Metrics

It includes prediction accuracy, network efficiency, and computational performance. The metrics include prediction accuracy metrics such as root mean square error (RMSE), mean absolute error (MAE), classification accuracy/F1-score, etc.

6.8.3.1 Network Performance Metrics

1. Spectrum Utilization Efficiency (SUE): It indicates the fraction of available channels successfully used for transmission.
2. Collision Probability: It measures the frequency of simultaneous transmissions.
3. Throughput: It represents average successful data rate per device or network segment.
4. Latency: It quantifies the delay between spectrum prediction and action

6.8.3.2 Computational Efficiency Metrics

The typical efficiency metrics include training time, inference time, and memory usage.

6.8.4 Blockchain-Enabled Spectrum Management for Secure and Transparent IoT Networks

Rapid growth of ultra-dense IoT networks has created a strong need for safer and decentralized spectrum management systems. Blockchain technology will provide an additional platform for creating spectrum intelligence systems in this field. Distributed ledger technology such as blockchain will keep a transaction record in a safer and tamper proof by sharing it across the multiple nodes.

6.8.5 Dataset Generation and Pre-Processing

Accurate simulation requires realistic datasets that capture the temporal, spatial, and spectral diversity of IoT networks. Key factors to consider include temporal Patterns, spatial characteristics, channel parameters, etc.

Typical pre-processing techniques involve normalization, feature extraction and data augmentation.

6.8.6 Comparative Evaluation of ML Models

Simulation studies allow systematic comparison of different ML approaches for spectrum forecasting, summarized in Table 6.5:

6.8.7 Case Study Example

To demonstrate this, consider an IoT network in a smart city with the following configuration:

- A 5 km^2 urban area covered by 2,000 sensors.
- Spectrum bands with several channels, each with different occupancy and interference levels.
- Devices that generate a combination of event-driven and periodic traffic patterns.

Simulation Configuration: NS3 integrates the LSTM and Deep Q-Network (DQN) models based on Python. Multi-agent reinforcement learning (MARL) is utilized for ensuring compliance with changing rules.

TABLE 6.5 Comparison of Different ML Approaches for Spectrum Forecasting

Model	Accuracy	Inference Latency	Scalability	Noise Robustness	Computational Efficiency	Interpretability
Linear Regression	Moderate	Low	High	Low	Very High	High
SVR	High	Moderate	Medium	Medium	Medium	Medium
Random Forest	High	Moderate	Medium	High	Medium	Medium
Gradient Boosting	Very High	Moderate	Medium	High	Low-Medium	Low-Medium
ARIMA	Moderate	Low	High	Low	High	High
LSTM/GRU	Very High	Moderate	Medium	High	Low-Medium	Low
CNN/ CNN-LSTM	Very High	High	Medium	High	Low	Low
Q-Learning / DQN	Medium-High	Moderate	Low Medium	Medium	Medium	Low
MARL	High	Moderate	Medium	High	Medium	Low
Federated Learning	High	Moderate	High	High	Medium	Low

To guarantee data privacy and effective coordination, edge nodes utilize federated aggregation.

For spectrum prediction, the CNN-LSTM + MARL model achieved 92% accuracy. The communication overhead is reduced by 45% with Federated edge learning. As compared to random access methods RL-based channel selection decreased collision probability by 35%. These results clearly show how the integration of ML, RL, and distributed learning frameworks can significantly optimize spectrum forecasting and access in complex IoT networks.

6.9 CHALLENGES, OPEN ISSUES, AND FUTURE DIRECTIONS

6.9.1 Introduction

These environments are specially marked by the massive device density, heterogeneous traffic, mobility, and highly dynamic spectrum conditions. There are few challenges regarding privacy, computational efficiency and regulatory compliance, when it comes to embedding AI and ML into real world spectrum management.

6.9.2 Key Challenges in Spectrum Forecasting for Ultra-Dense IoT

6.9.2.1 Data Scarcity and Quality

To impart accurate predictions, the machine learning models need massive and high-quality datasets. Many IoT devices usually produce sparse and intermittent data, limiting the performance of supervised models. The prediction accuracy is further reduced by interference, noise incurrence and sensor errors. To address the challenges related to incomplete and noisy spectrum measurements, the semi-supervised learning techniques, robust preprocessing, and data augmentation techniques are essential.

6.9.2.2 High Dimensionality and Heterogeneity

Multiple dimensions like time, location, frequency, and device types are involved in IoT spectrum data. Non-uniform data distributions are incurred due to the devices operating using different protocols, transmission powers, and traffic characteristics. The prime requirements of effective forecasting are feature extraction, dimensionality reduction, and deep learning architectures capable of managing diverse data sources.

6.9.2.3 *Computational and Energy Constraints*

There are constraints in terms of CPU, memory, and battery capacity for many IoT nodes. It is always challenging for the devices when it comes to running complex models like LSTM, CNN, or multi-agent RL. For real world deployment, it is necessary to have lightweight ML architectures, edge-assisted processing, and model compression techniques.

6.9.2.4 *Dynamic and Non-Stationary Environments*

The characteristics such as mobility, new devices, policy changes, and interference affect the spectrum usage. These variations can be easily adaptable using Static ML models trained offline. Continuous learning, online adaptation, and reinforcement learning are required for maintaining prediction accuracy in changing environments.

6.9.2.5 *Privacy, Security, and Regulatory Compliance*

Centralized data collecting increases the possibility of issues in spectrum usage and may affect the privacy regulations. Privacy-aware frameworks such as secure aggregation, FL, and differential privacy must be addressed efficiently to guarantee compliance with changing rules.

6.9.3 Open Issues in Current Research

1. Multi-Agent Coordination: MARL improves distributed decision-making, but becomes challenging to achieve scalability, stability, and fairness.
2. Real-Time Spectrum Intelligence: Low latency forecasting affects the industrial automation, smart grids, and driverless cars.
3. Model Generalization Across Regions: There exists geographical and temporal variations in spectrum patterns.
4. Explainability and Interpretability: Explainable AI (XAI) helps in maintaining accountability, fostering confidence, and adhering to legal requirements.
5. Integration with Policy and Standardization: ML-based spectrum management needs to be aligned with rules and regulations that includes DSA standards and 5G NR unlicensed spectrum guidelines.

6.9.4 Future Directions

- Hybrid Machine Learning Frameworks: By integrating supervised, unsupervised, and reinforcement learning, their corresponding benefits can be utilized.
- Edge-Cloud Collaborative Learning: By utilizing a joint edge cloud design, there can be a descent balance between local responsiveness and global optimization.
- Transfer Learning and Domain Adaptation: By using pre trained models, the deployment in new regions can be accelerated.
- AI-Driven Spectrum Policy Management: Spectrum negotiation and policy compliance can be automated by combining machine learning (ML) and CR systems.
- Lightweight and Energy-Efficient Machine Learning Models: By using methods like knowledge distillation, quantization, and model pruning, the computational demands can be reduced.
- Explainable and Trustworthy AI: Explainable AI (XAI) approaches can make spectrum decisions transparent to network administrators.

CONCLUSION

In ultra-dense IoT system, an efficient and autonomous spectrum management is provided by machine learning integrated spectrum forecasting. This study provides some useful information for practitioners utilizing spectrum intelligent IoT systems: Coordination of technological and policy initiatives is necessary for the successful deployment of ML-driven spectrum forecasts in extremely dense IoT contexts.

To eliminate centralized data gathering, FL is advised for privacy-sensitive applications. To support mission-critical IoT applications, real-time spectrum decisions should achieve latency targets of less than 100 ms. For the aid of ML Researchers, to guarantee model robustness and generalization, create a variety of labeled datasets from indoor, outdoor, and urban IoT contexts. To verify performance improvements, always compare deep learning techniques to traditional baselines like ARIMA. To lower the amount of data needed in new deployment scenarios, look at transfer learning strategies.

Before widespread deployment, spectrum regulators should start ML-driven DSA pilots in unlicensed bands. The regulatory frameworks must be updated regularly to enable formal acceptance of AI-based channel selection techniques. IoT experts suggest deploying pre-trained models and fine-tuning them along with local spectrum data typically also use neural network quantization (e.g., 8-bit) for embedded devices.

In the domains of explainable AI, domain adaptation, federated and edge assisted learning, hybrid machine learning frameworks, and lightweight architectures it is necessary to decide the priorities of future advancements. For future IoT contexts, these discoveries may result in reliable, independent, and scalable spectrum management, particularly when combined with collaborative intelligence and flexible regulations. Spectrum forecasting can transit from a laboratory setting to practical environment by efficiently resolving these issues and incorporating cutting-edge AI technology. This drives efficiency, fairness and resilience across smart, connected setups.

PRACTITIONERS TAKEAWAY

In context of ultra-dense IoT setups, the spectrum forecasting based on machine learning provides a well-defined solution to improve the utilization of spectrum, reduce the interference, and provide a reliable communication medium. It is necessary to choose the prediction models by practitioners by considering the deployment limitations including data availability, accessibility, delay necessities, processing power, and privacy issues. Usually, deep learning and reinforcement learning models offer higher accuracy for large and dynamic networks. The lightweight models such as random forest and SVR are suitable for edge devices. Federated and edge assisted learning frameworks play an important role to provide distributed intelligence and privacy protection. It is very important to prepare appropriate dataset and focus on performance measurements using network-level metrics, and compliance with regulatory standards in order to guarantee scalable and reliable spectrum management systems.

REFERENCES

[1] M. G. S. Murshed, C. Murphy, and T. K. Das, "Machine Learning at the Network Edge: A Survey," *ACM Computing Surveys*, vol. 54, no. 8, Art. 177, pp: 1-37, 2021. DOI: 10.1145/3469029.

[2] F. R. V. Guimarães, D. B. da Costa, and M. Cavalcante, "Machine Learning for Spectrum Sharing: A Survey," *Foundations and Trends in Networking*, vol. 16, no. 3–4, pp. 207–347, 2024. DOI: 10.1561/1300000037

[3] M. S. Mahdavinejad, M. Rezvan, M. B. Phung, and P. Azimi, "Machine Learning for Internet-of-Things Data Analysis: A Survey," *Digital Communications and Networks*, vol. 4, no. 3, pp. 161–175, 2018. DOI: 10.1016/j.dcan.2017.10.002.

[4] M. Kulin, T. Kazaz, E. De Poorter, and I. Moerman, "A Survey on Machine Learning-Based Performance Improvement of Wireless Networks: PHY, MAC and Network Layer," *Electronics*, vol. 10, no. 3, Art. 318, 2021. DOI: 10.3390/electronics10030318.

[5] Waseem, Q., Din, W.I.S.B.W., Rahman, A.B.A. *et al.* A survey and analysis of feature selection techniques in machine learning for IoT device classification within smart buildings. *Innov. Infrastruct. Solut.* 10, 407 (2025). https://doi.org/10.1007/s41062-025-02203-7

[6] M. S. Rahman, S. Begum, M. Alam, and M. F. Hossain, "Machine Learning and Internet of Things in Industry 4.0: Challenges and Applications," *Computers and Industrial Engineering*, vol. 188, Art. 109929, 2023. DOI: 10.1016/j.cie.2023.109929.

[7] J. Liu, X. Zhou, and M. Bennis, "Machine Learning-Driven 6G Wireless Networks: Architecture, Use Cases, and Challenges," *IEEE Communications Surveys & Tutorials*, vol. 24, no. 4, pp. 2431–2472, 2022. DOI: 10.1109/COMST.2022.3172490.

[8] H. Song, L. Liu, J. Ashdown, and Y. Yi, "A Deep Reinforcement Learning Framework for Spectrum Management in Dynamic Spectrum Access," *IEEE Internet of Things Journal*, vol. 8, no. 14, pp. 11208–11218, 2021. DOI: 10.1109/JIOT.2021.3052691.

[9] F. Li, B. Moon, and Z. Cao, "A Dynamic Spectrum Access Scheme for IoT Based on Federated Deep Reinforcement Learning," *Journal of Network and Computer Applications*, vol. 250, pp. 7952–7956, Art. 104189, 2025. DOI: 10.1016/j.jnca.2025.104189.

[10] R. Safavinejad, M. J. Salehi, M. Shikh-Bahaei, and H. Arslan, "Deep Reinforcement Learning for Dynamic Spectrum Access: Convergence Analysis and System Design," *IEEE Transactions on Wireless Communications*, vol. 23, no. 12, pp. 18888–18902 , 2024. DOI: 10.1109/TWC.2024.3414428.

[11] A. Kaur and S. K. Aseri, "A Comprehensive Survey on Machine Learning Approaches for Dynamic Spectrum Access in Cognitive Radio Networks," *Journal of Engineering and Technology*, *34*(1), 1–40, 2022. DOI: 10.1080/0952813X.2020.1818291.

[12] O. Naparstek and K. Cohen, "Deep Multi-User Reinforcement Learning for Distributed Dynamic Spectrum Access," *IEEE Transactions on Wireless Communications*, vol. 18, no. 1, pp. 310–323, 2019. DOI: 10.1109/TWC.2018.2879433.

[13] F. Li, Y. Wang, and J. Xu, "Dynamic Spectrum Access for Internet-of-Things with Joint GNN-DQN Optimization," *Journal of Network and Computer Applications*, vol. 238, Art. 103789, 2024. DOI: 10.1016/j.jnca.2024.103789.

[14] F. Li, B. Moon, J. Kim, and K. H. Chang, "Dynamic Spectrum Access for IoT Services in Cellular and Unlicensed Bands," *Computer Networks*, vol. 133, pp. 40–54, 2017. DOI: 10.1016/j.comnet.2017.03.004.

[15] A. Ahmed, M. M. Hassan, and C. Xu, "Federated Learning-Assisted Edge Intelligence for Dynamic Spectrum Allocation in 6G Networks," *IEEE Network*, vol. 37, no. 4, pp. 167–173, 2023. DOI: 10.1109/MNET.2023.3274695.

[16] D. Jaramillo-Ramírez, "Spectrum Demand Forecasting for IoT Services," *Future Internet*, vol. 13, no. 9, Art. 232, pp. 1–18, 2021. DOI: 10.3390/fi13090232.

[17] C. Brown, E. Singh, and A. Lopez, "Enhanced IoT Spectrum Utilisation: Integrating Geospatial and Real-Time Data for Adaptive Allocation," *Sensors*, vol. 24, no. 18, Art. 5885, 2024. DOI: 10.3390/s24185885.

[18] Z. Qin, X. Zhou, and G. Y. Li, "6G-Enabled Edge Intelligence for Spectrum Management in IoT Networks," *IEEE Wireless Communications*, vol. 30, no. 1, pp. 62–69, 2023. DOI: 10.1109/MWC.2023.3221123.

[19] Y. Zhang, L. Qian, and D. K. C. So, "Deep Learning Empowered Spectrum Prediction for Cognitive IoT Networks," *IEEE Transactions on Cognitive Communications and Networking*, vol. 8, no. 2, pp. 510–523, 2022. DOI: 10.1109/TCCN.2021.3138279.

[20] A. K. Bashir, R. Qureshi, and H. S. Kim, "AI-Based Spectrum Forecasting for Efficient Cognitive IoT Communication," *IEEE Access*, vol. 10, pp. 106453–106468, 2022. DOI: 10.1109/ACCESS.2022.3204911.

[21] M. Elsayed, M. Erol-Kantarci, and A. Mohamed, "AI and Deep Learning for Spectrum Management in 6G Networks," *IEEE Internet of Things Magazine*, vol. 6, no. 2, pp. 92–97, 2023. DOI: 10.1109/IOTM.2023.3235561.

[22] R. Shafin, L. Liu, and Y. Yang, "Artificial Intelligence-Enabled Cellular Networks: A Critical Path to 6G," *IEEE Communications Magazine*, vol. 59, no. 2, pp. 76–82, 2021. DOI: 10.1109/MCOM.001.2000542.

[23] M. A. Aygül, M. Nazzal, M. İ. Sağlam, D. B. da Costa, H. F. Ateş, and H. Arslan, "Efficient Spectrum Occupancy Prediction Exploiting Multidimensional Correlations Through Composite 2D-LSTM Models," *Sensors*, vol. 21, no. 1, Art. 135, 2021. DOI: 10.3390/s21010135.

[24] M. A. Aygül, M. Nazzal, A. R. Ekti, A. Görçin, D. B. da Costa, H. F. Ateş, and H. Arslan, "Spectrum Occupancy Prediction Exploiting

Time and Frequency Correlations Through 2D-LSTM," in *Proc. IEEE Vehicular Technology Conference (VTC-Spring)*, Antwerp, Belgium, 2020. DOI: 10.1109/VTC2020-Spring48590.2020.9129001.

[25] A. Ghosh, S. Banerjee, M. Roy, and A. Mukherjee, "Spectrum Usage Analysis and Prediction Using Long Short-Term Memory," in *Proc. ACM/IEEE International Conference on Computing and Communication Systems*, 2023. DOI: 10.1145/3571306.3571412.

[26] N. Zhao, X. Pang, and Y. Chen, "Energy-Efficient Spectrum Prediction for Cognitive Wireless Networks Using Deep Learning," *IEEE Transactions on Green Communications and Networking*, vol. 6, no. 2, pp. 840–854, 2022. DOI: 10.1109/TGCN.2022.3150523.

[27] H. Yang, J. Li, T. Q. S. Quek, and C. Xu, "Lead Federated Neuromorphic Learning for Wireless Edge," *Nature Communications*, vol. 13, Art. 4578, 2022. DOI: 10.1038/s41467-022-32020-w.

[28] Q. Li *et al.*, "A Survey on Federated Learning Systems: Vision, Hype and Reality for Data Privacy and Protection," in *IEEE Transactions on Knowledge and Data Engineering*, vol. 35, no. 4, pp. 3347-3366, 1 April 2023, doi: 10.1109/TKDE.2021.3124599.

[29] F. Tang, M. Kawano, and K. Ota, "Federated Learning-Based Distributed Spectrum Sensing for Cognitive IoT," *IEEE Internet of Things Journal*, vol. 10, no. 1, pp. 255–265, 2023. DOI: 10.1109/JIOT.2022.3211829.

[30] K. B. Letaief, W. Chen, and Y. Shi, "Edge Intelligence for 6G-Enabled Internet of Everything: Architecture and Key Technologies," *Proceedings of the IEEE*, vol. 110, no. 8, pp. 1251–1272, 2022. DOI: 10.1109/JPROC.2022.3172197.

CHAPTER 7

Blockchain-Enabled Spectrum Auctions and Regulatory Frameworks

Archana Rajesh Date, Anuradha Nigade, Zaid Alsalami, and Anurag Shrivastava

7.1 EVOLUTION OF SPECTRUM MANAGEMENT AND REGULATORY PARADIGMS

The radio frequency spectrum is becoming one of the most precious and scarcest resources on earth due to the rapid growth of wireless communications technology. In order to maximize industry's innovative potential, minimize coexistence problems, and optimize the resource, radio frequency spectrum management is essential [1]. From administratively required and normally managed monitoring to a law that is concerned with technology, spectrum management has changed over time. Changes in worldwide collaboration, policy agendas, and communications demands have led to this transformation. A systematic study of support and management of the radio frequency spectrum and its regulatory conditions increases understanding of how governments and organizations maintain order for the invisible flow of radio waves.

7.1.1 Early Days: The Radio Spectrum and the Development of Regulation (Pre-20th Century Until the Early Part of the 20th Century)

Background: The idea of radio communication began in the late 1800s with early pioneers like Marconi demonstrating wireless transmission. At that

DOI: 10.1201/9781003731597-7

time, however, radio waves were unregulated, leading to many instances of interference from other early radio operators.

Need for Regulation: As radio communication expanded, especially in maritime communication, issues of interference arose whereby governments had to step in to help regulate the airwaves. In first Regulatory Action an International Telegraph Union (ITU) was formed in 1865 to act as a coordination tool for telegraph communication. In the early 1900s, the ITU's role was extended to include radio communication and they began working toward coordinating the radio spectrum across the globe.

7.1.2 1920s–1930s: The Emergence of Formal Spectrum Management

1927–The U.S. Federal Radio Commission (FRC): The United States created the FRC (became the FCC in 1934) to regulate radio frequencies and technical standards in an effort to limit interference. The ITU held a series of international Radiotelegraph Conferences to allocate frequencies and services internationally. The radio spectrum was beginning to be allocated depending on service (broadcasting, maritime, military, etc.) and geography.

7.1.3 Post-WWII Era and the Expansion of Telecommunications (1940s–1960s)

Establishment of Radio Regulations and ITU Constitution (1947): The ITU implemented Radio Regulations, an international treaty that established framework for usage of frequencies globally, established technical norms and procedures for coordinating frequency usage.

Growth of Services: New uses, such as broadcasting TV, business aviation, and the first mobile radio demanded for additional spectrum. Spectrum management was largely controlled; governments issued exclusive rights to operators and assigned technical specifications in detail to avoid interference. National regulatory agencies developed long term plans for spectrum usage balancing military, commercial and public safety protection.

7.1.4 1970s–1980s: Increased Technological Innovation and Demand for Spectrum Usage Growth

Technological Change: The arrival of satellite communications, cellular radio (1G), and the advancement of spread spectrum technologies increased demand for spectrum and technical complexity. A reliance on rigid licensing policies resulted in inefficient use of spectrum in some bands and congestion in others. Some countries, e.g., New Zealand in 1989, introduced auctions to allocate spectrum licenses as a way to efficiently manage those spectrum/licenses, either as a substitute for the administrative process, contemporary with it. The auctions defined technical standards (power, emission masks), coordination procedures and criteria for rights of protection against interference.

7.1.5 1990s–2000s: Liberalization and Market Mechanisms

A Move to Market Conceptions: Spectrum trading and the establishment of secondary markets enabled license holders to buy, sell, or rent spectrum rights. Regulators started to allow more flexible uses of assigned spectrum to incentivize innovation and neutrality of technologies. The rise of the ideas of cognitive radio and dynamic spectrum access: Research in radios that could sense and adapt to spectrum availability began, introducing complexities to the prior static allocation [6]. International cooperation broadened, particularly to promote mobile communications (e.g., GSM, 3G).

7.1.6 2010s–Present: Dynamic Spectrum Management and Convergence

Dynamic Spectrum Access and Sharing: Technologies such as TV White Spaces, Citizens Broadband Radio Service (CBRS) in the U.S. highly advocate for the more dynamic sharing of spectrum.

Auctions and Pricing Models: Diverse models for auctions of spectrum and pricing models are used including multi-round, reserve prices and specified caps in spectrum, to ensure balance between competition and capturing revenue.

7.2 IMPORTANT REGULATORY FACTORS IN SPECTRUM MANAGEMENT

Frequency Allocation and Assignments:

- Allocation: Partitions within the spectrum where it is expected that specific services can utilize the spectrum (mobile, fixed, broadcast).
- Assignment: Licenses or rights to use parts of the spectrum within the allocated frequency bands.

License Models:

- Exclusive License: The complete ability to use (an) assigned (allocations of frequency from the frequency allocation) spectrum in a given geographic area.
- Shared License: Spectrum can be utilized by multiple users (unlicensed bands, for example, Wi-Fi).
- Spectrum Commons: Open-access bands used which usually contain limitations to prevent harmful interference from occurring.

Technical Factors:

- Power Limits: Maximum transmission power that can be transmitted that limit the range of interference.
- Emission Masks: Limitations on out-of-band emissions to limit interference with the assigned bands.
- Channel Bandwidth and Modulation: Specific definitions that ensure parameters fall within reasonable constraints, ensuring compatibility and efficiency.

Interference Management:

- Coordination and Notification: Procedures to avoid and manage interference when working across borders and across services.
- Protection Criteria: Define tolerance thresholds to avoid interference.

Economic and Administrative Factors:

- Fees and Charges: Licensing fees and spectrum usage charges.
- Auctions and Market Rules: Defined procedures for competitive allocation of licenses to services and users.
- Secondary Markets: Define the rules governing the buying/selling and leasing of allocated and unallocated frequencies.

7.2.1 Mechanisms for Auctions of Spectrum

The economics of auctions for spectrum have a transformative role in the way governments allocate licenses for spectrum. The notion is to better allocate spectrum to the highest value in an efficient, transparent, fair, and profitable manner for the government.

a. *Types of Auctions:*

First-Price Sealed-Bid Auction: All bidders submit their final bids in sealed envelopes. The highest bidder is the winner and pays the bid price. This is simple to understand, though it often encourages bidding behavior that involves strategy and possibly underbidding.

Second-Price Sealed-Bid Auction (Vickrey auction): The highest bidder is the winner but pays only the second highest bid. Thus, it encourages bidders to expose their true valuation.

Simultaneous Multiple-Round Auction (SMRA): Bidders can bid on multiple licenses simultaneously across a series of rounds; this provides an opportunity for bidders to revise their bids based on information on rival's bidding.

Combinatorial Auctions: Bidders can bid on combinations of spectrum blocks (packages); such auctions are useful when the blocks of spectrum have interdependent of the blocks or packages on the total valuation of the signals.

Clock Auctions: In a clock auction, prices increase over a series of rounds ("clock ticks") and the bidders then have an opportunity to indicate how much quantity they want at the pricing at the ending of each clock tick or round. The rounds continue until there is equality in demand and supply.

7.2.2 International Treaties and Coordination

Role of ITU:

The ITU coordinates worldwide spectrum usage and defines spectrum use for numerous services of architectural interest. ITU coordination is governed by the Radio Regulations (RR) treaty that is negotiated every 3-4 years during the World Radiocommunication Conferences (WRC).

The RR specifies: International frequency allocations, protocols for sharing and coordinating the use of frequencies, technical specifications for terrestrial and satellite systems, protection regulations for services including broadcasting, satellite, and aviation.

Coordination Procedures:

- Advance Publication and Notification: Countries provide notices to the ITU and neighboring countries about the intended use of spectrum in the coordination area.

- Bilateral/Multilateral Agreements: Neighboring countries agree to methods to mitigate potential interference.
- Dispute Resolution: The ITU serves as an arbitrator to resolve conflicts.

7.2.3 Dynamic Spectrum Sharing and Emerging Paradigms Dynamic Spectrum Access (DSA)

Spectrum is planned and accessed dynamically based on real-time availability of the resource.

a. *Techniques include:*

   - Cognitive Radio: Radios that sense the spectrum environment to opportunistically access unused licensed spectrum.
   - TV White Spaces (TVWS): The unused spectrum range of TV broadcast is available for secondary use provided that the secondary usage does not cause interference to the incumbent broadcasters.
   - Licensed Shared Access (LSA): Users who are authorized to access and share the resource shared frequencies with the incumbent usage, under community conditions.
   - Citizen Broadband Radio Service (CBRS): The United States employs a three-tiered framework for accessing and dynamically using the 3.5 GHz band.

b. *Benefits of dynamic sharing:* Improved efficiency by sharing the spectrum., Makes available spectrum accommodations for more users and services., Reducing barriers for entry for new operators.

c. *Regulatory Challenges:* Identifying methods to define rules for interference protection. developing reliable sensing or sensing database technology. Ensuring fairness amongst priority of access.

7.3 5G SPECTRUM

The 5G spectrum refers to the range of radio frequencies allocated for fifth-generation (5G) mobile communication technology. It has a significant impact on 5G networks' capacity, speed, and coverage. 5G offers many forms of connectivity by combining the low, mid, and high frequency bands. Wide coverage but slower speed is provided by low-band (below

1 GHz). Most commonly used for 5G, mid-band (1–6 GHz) achieves a balance between speed and coverage. High-band (mmWave, above 24 GHz) offers extremely fast speeds and minimal latency, although its coverage's constrained. Because the 5G spectrum offers high bandwidth, quicker data rates, and dependable connections, it makes applications like the Internet of Things, smart cities, driverless cars, and real-time communication possible. Efficient spectrum management and allocation by regulatory bodies (like TRAI in India or FCC in the USA) are essential for smooth operation of 5G networks. Technical Standards in 5G Spectrum

7.3.1 Key Players

- 3GPP (3rd Generation Partnership Project): Responsible for developing radio interface standards for 5G (NR – New Radio).
- Frequency range 1 (FR1) is the sub-6 GHz frequencies, such as 3.3–3.8 GHz and 4.4–5.0 GHz bands.
- Frequency range 2 (FR2) is the millimeter wave (mmWave) bands; frequency ranges 24.25–27.5 GHz, 37–43.5 GHz, and 45.5–47 GHz.

These two frequency ranges are important to the high-speed, low-latency features of the 5G wireless technology.

- ITU (International Telecommunication Union): Establishes global spectrum allocations and framework through the WRC process.
- IEEE: Contributes to some supporting complementary wireless standards (e.g., Wi-Fi 6/7). These standards are typically in adjacent bands or expressly in unlicensed bands.

Fundamental Technical Parameters of 5G NR (3GPP TS 38.101)

- Carrier Bandwidth:
- Up to 100 MHz in FR1 (and may accommodate 200MHz in future releases).
- Up to 400MHz (or wider) in FR2 with carrier aggregation.

Channel Bandwidth: Allows for flexible numerologies (subcarrier spacing of 15 kHz to 240 kHz), allowing for scalability and low-latency performance. Massive MIMO & Beam forming improves spectrum efficiency, particularly in mm Wave bands. Dynamic spectrum sharing (DSS) allows for 4G LTE and 5G NR to co-exist dynamically in the same spectrum bands.

7.3.2 WRC Decisions Relevant to 5G Spectrum

WRC is convened by the ITU (International Telecommunications Union) every three to four years to study and approve Radio Regulations (international treaty regarding use of spectrum).WRC decisions provide or assign spectrum bands for specific service, such as 5G and mobile broadband.

WRC Decisions Significance:

- Global consistency: WRC decisions aid manufacturers in developing equipment that is compatible around the world, thus lowering costs.
- Spectrum coordination: WRC decisions reduce the possibility of interference across national borders or prevent it altogether and help public-sector agencies schedule auctions.
- Roadmaps of technology: Working to ensure that spectrum availability is synchronized with technology standards (sometimes on an international basis) like releases from 3GPP to alleviate any negative impacts when wireless technology is launched into service.

These decisions are part of international efforts to address continued demand for additional wireless communications services and to utilize spectrum efficiently.

[4]

7.3.3 Key Challenges in 5G/6G Era

7.3.3.1 Spectrum Fragmentation

The partition of the accessible radio spectrum into numerous tiny, non-contiguous frequency blocks (pieces) is known as fragmentation. To put it simply, operators get small, dispersed parts across many frequency bands rather than a single, large, continuous block of spectrum. Spectrum Fragmentation Causes:

- Many allocations throughout time: Spectrum is not released entirely at once, but rather in phases.
- Different technologies: The frequency ranges used by 2G, 3G, 4G, and 5G are all different.
- Multiple operators: Numerous telecom firms share spectrum.

TABLE 7.1 Timeline of WRC Decisions for 5G/6G Spectrum[4]

Year	WRC Conference	Key Decisions and Highlights	Impact and Notes
2012	WRC-12	Identified the need to study bands for future IMT (International Mobile Telecommunications) systems.	Set groundwork for 5G spectrum discussions; early recognition of mm Wave potential.
2015	WRC-15	- Officially allocated 3.3-3.4 GHz and 3.4-3.6 GHz bands for IMT worldwide. – Candidate bands 4.4-5.0 GHz for IMT identified. – 24.25-27.5 GHz band identified for future IMT.	Major milestone: mid-band and early mm Wave bands recognized globally for 5G. Enabled initial 5G deployments.
2019	WRC-19	- Extended mm Wave spectrum allocations to 37-43.5 GHz, 45.5-47 GHz.- Added 6425-7125 MHz mid-band for IMT.- Confirmed 24.25-27.5 GHz for 5G use.- Studies initiated on 66-71 GHz.	Supported ultra-high-capacity 5G use cases. Helped harmonize mm Wave bands internationally. Broadened mid-band options.
2023	WRC-23	- Identified 6425-7125 MHz band formally for IMT. – Considered bands above 52.6 GHz for 6G and beyond. – Discussions on sharing mechanisms for bands like 10-10.5 GHz with incumbent satellite services. – Focus on dynamic spectrum sharing and coexistence.	Key step toward enabling 6G spectrum. Enhanced mid-band availability for 5G expansion. Addressed coexistence challenges critical for future networks.

- Mergers and acquisitions: Businesses trade or merge, forming irregular blocks.

7.3.3.2 Impacts of Spectrum Fragmentation

- Increased Complexity in Device Design: Devices need to support so many bands, with complex antennas, creating added cost and power consumption.
- Lower Efficiency of Spectrum Use: Fragmented spectrum reduces the capacity to aggregate spectrum use to certain x dates and limits the total amount of bandwidth used per user.
- Complications in Aggregating Spectrum: Carrier aggregation (combining spectrum to get better throughput) creates more complications while spectrum is fragmented.

- Interoperability Constraints: Fragmented spectrum creates a barrier to seamless roaming and cross-border communications.
- Potential Solutions:
  - o Spectrum Reframing: Reassigning and restructuring spectrum from previous services to modern technologies
  - o Flexible Spectrum Management and Sharing: Allowing access to dynamically use spectrum and fragmented bands.
  - o Harmonization initiatives: Collaboration on a worldwide scale to harmonize the band for 5G/6G.

7.3.3.3 Scarcity of Spectrum

Scarcity of spectrum means having a finite amount of usable radio frequencies, particularly those in scarcity spectrum (i.e., scarcity-driven spectrum division) that cannot accommodate the increasing demand for data-heavy 5G/6G services.

Reasons for Scarcity:

- Exponential Growth in Wireless Data: The requirement for 5G/6G will allow massive IoT, ultra-HD video, AR/VR, autonomous vehicles-- all of which require enormous bandwidth.
- Limited Low/Mid-Band Spectrum: Limited low-mid band is the type of spectrum that works best for coverage and penetration. Thus, they are becoming congested with the legacy user base.
- Exclusivity to Spectrum: Licensing spectrum based on past models grants exclusivity for use, which also encourages under use and "spectrum hoarding."

7.3.3.4 Impacts of Scarcity

- High Spectrum Costs: Costs to obtain spectrum via auction become extremely high and thus a barrier to entry for newcomers.
- Networking Congestion: Spectrum that is not available and affects costs also leads to slower speeds or reduced quality of service.
- Limits Innovation: Scarcity of spectrum will also curb any deployment of new service availability and applications, including those that are developing.

7.3.3.5 Potential Solutions

- High Band Usage (faster than 4G): More usage of millimeter-wave (mm Wave) and possibly sub-THz for access in large bandwidth at lower than 4G data rates, but with limited levels of coverage and penetration in terms of radio capacity.
- Spectrum Sharing Models: Network models that leverage Dynamic sharing model, Licensed Shared Access (LSA), some unlicensed spectrum access possibly.
- Technologies for Better Spectrum Efficiency: Massive MIMO, highly precoded beam forming, or new, more precise types of coding that can deliver more spectrum usage limitations to current spectrum portfolios.
- Regulatory Reforms: Support spectrum trading, secondary markets, and other incentives for efficient spectrum use.

7.3.3.6 Complexity in Spectrum Management

Complexity derives from the complicated technical, regulatory, and operational challenges that result from new technologies, the breadth of use cases, and global coordination to manage multiple service providers.

7.3.3.7 Causes of Complexity

- Multiple Frequency Bands and Technologies: 5G/6G networks approach their networks with a combination of different low, mid, and high bands with different propagation characteristics.
- Diverse Use Cases: eMBB (enhanced mobile broadband), URLLC (ultra-reliable low latency communications) [8]and mMTC (massive machine-type communications) all have very different requirements.
- Dynamic and Shared Spectrum Access: New paradigms (CBRS, TV White Spaces, etc.) introduce the need for real-time coordination and interference management for spectrum access.
- Heterogeneous Networks (HetNets): There is an increasing integration of macro cells, small cells, Wi-Fi, and IoT networks that will alter interference and resource management.
- Cross-Border Coordination: Using spectrum near borders requires intricate agreements and coordination to manage potential interference.

- Security Concerns: Increased devices and spectrum sharing create new vulnerabilities and require maximizing security protocols[3]

7.3.3.8 Impacts of Complexity

- Operational Challenges: Regulators and operators require advanced tools to monitor spectrum use and impacts from other services.
- Increased Costs: Complexity will drive up costs for R&D, deployment and operational expenditures.
- Slow Policy Adaptation: At times the challenge is the inability of the regulatory document to shift with all these technology changes in the landscape.

7.3.3.9 Possible Solutions

- Enhanced Spectrum Management Tools: AI and machine learning (ML) can enable dynamic spectrum allocation, interference detection, and predictive analytics.
- Policy Innovation: Agile and adaptive regulatory frameworks can enable rapid technology evolution.
- Global Collaboration: More effective coordination in international standardization efforts.
- Automation and Cognitive Radio: Automation of spectrum sensing and management would help to decrease reliance on manual handling.

7.4 REGULATORY APPROACHES TO SOLVING SPECTRUM PROBLEMS

7.4.1 Spectrum Reframing and Reallocation

- Regulators recognize the legacy spectrum that is not being used enough (i.e., 2G/3G frequency bands) and reassign it for 5G/6G.
- Allows for more contiguous spectrum to be used for new technologies.

7.4.2 Encouraging Spectrum Sharing Policies

- Develop clear frameworks for various types of sharing models (e.g., LSA/CBRS).
- Develop more formalized licensing approaches and secondary markets for trading or leasing spectrum rights.

- Encourage use of unlicensed spectrum but outline obligations and protocols for managing interference.

7.4.3 Harmonization and Global Coordination

- Coordinated spectrum allocation across countries would allow for increased economies of scale for manufacturers of devices.
- Coordination and standardization would take place through the ITU and regional bodies to develop common bands and technical standards.

7.4.4 Incentivizing Efficient Use of Spectrum

- "Use-it-or-share-it" policies in which licensees must either deploy there services or lose their rights to the spectrum.
- Encourage technologies that allow for increased spectral efficiency of devices and services.
- A model for pricing spectrum that may reflect demand but also encourages efficiency/efficient use.

7.4.5 Dynamic Licensing and Flexible Use Rights

- Moving away from the rigid, long term exclusive licensing for spectrum to dynamic or flexible licenses that can adjust to changing licensing needs.
- Licenses based on temporary use or geographic/base use (e.g., for special events or rural coverage).

7.5 SPECTRUM ACCESS SYSTEM (SAS)

A spectrum access system (SAS) is a spectrum management system that functions in real time, allowing different priority users access to operate within a shared band, furnishing "interference protection" for each incumbent user and optimizing overall spectrum utilization. Step-by-step workflow is explained below.

7.5.1 SAS workflow

7.5.1.1 Registration

- Registration of CBSDs: Any user device (CBSD) wanting to transmit on the shared band must register with the SAS and share device characteristics (including location, antenna height, and device capabilities).

- License Verification: SAS determines whether the user is a PAL user or a GAA user.

7.5.1.2 Spectrum Inquiry and Grant Request

- Spectrum Inquiry: CBSD sends inquiry to SAS for available spectrum at location and requests permission to use specific channels.
- Grant Issuance: SAS assesses the present spectrum utilization, the existence of incumbents and their protection rules, and grants authority for specific frequencies, power levels and timelines.

7.5.1.3 Interference Management and Protection

- Incumbent Protection: ESC sensors continually monitor for incumbent signals. When an incumbent signal is detected, SAS instructs impacted CBSDs to cease operation or reduce power on those frequencies.
- Priority Enforcement: PAL users are granted priority access, GAA users can only use channels that have not been assigned to incumbents or PALs.
- Dynamic Reassignment: AS dynamically reallocates spectrum when users enter or leave or when an incumbent becomes active or inactive.

7.5.1.4 Monitoring and Enforcement

- Compliance Monitoring: The SAS performs compliance monitoring of CBSD transmissions via either direct reports from CBSDs or sensing.
- Violation Handling: If a device is determined to be transmitting outside of its grant parameters, SAS may revoke access or take steps to limit permissions.

Data Logging and Reporting

- SAS has logs of all spectrum use, grants issued, interference events, and enforcement actions for auditing and/or regulatory compliance purposes.

7.5.2 Benefits of Dynamic Spectrum Sharing

- Maximizes the efficient use of spectrum by dynamically sharing it with other users.
- Protects critical incumbent services with real-time sensing and coordination.
- Supports multiple and diverse classes of users with different priority designations and levels.
- Eases the near-term deployment of 5G and private networks without the need for an exclusive spectrum license.

7.6 AI, ML, AND BLOCKCHAIN IN TELECOM OVERVIEW

7.6.1 Artificial Intelligence (AI) in Telecommunications

AI encompasses systems or machines that emulate human intelligence to perform tasks such as decision-making, problem-solving, and learning.

AI Applications in Telecom:

- Network Optimization: AI algorithms can analyze data and network usage patterns to optimize routing and bandwidth allocation while dynamically reducing congestion[4].
- Predictive Maintenance: AI predicts failures that may occur in a network piece of equipment by analyzing historical data and relevant sensor readings, allowing proactive maintenance to mitigate downtime.
- Customer Service Automation: AI-enabled Chabot's and virtual assistants can tackle customer queries, complaints, and troubleshooting, 24/7 – leading to improvement in the user experience.
- Fraud Detection: AI detects "suspicious" behaviors or patterns in call records, and data usage, thus, indicating fraudulent activity.
- Network Security: AI detects anomalies and potential cyber-attacks, while analyzing network traffic in real-time.
- Service Personalization: AI uses profiling of user's behavior in order to offer curated content, service plans and advice on service.

Benefits: Improved efficiency and reliability, lower operational costs, improved customer experience, and faster decision execution.

7.6.2 Machine Learning (ML) for Telecommunications

An element of AI that involves algorithm training based on data, so that it can detect patterns and create decisions or predictions without being programmed to do so.

ML Techniques of telecom:

- Supervised Learning: Used for classification and prediction based tasks, such as churn prediction[10].
- Unsupervised Learning: Clustering customers or identifying unknown anomalies.
- Reinforcement Learning: Time-tested configurations that can optimize a network through trial-and- error feedback loops.

Key Use Cases:

- Network Traffic Forecasting: ML models will predict network load allowing for efficient allocation of resources before the network is taxed[5].
- Churn Prediction: ML will identify those customers more likely to terminate relationships so they can be targeted for retention efforts.
- Anomaly Detection: Identify abnormal behavior in the network that indicates potential faults or breaches
- Energy Efficiency: ML algorithms will optimize energy consumption in network equipment.
- Quality of Experience (QoE) Enhancement: Predict parameters changes to ensure quality service delivery.

Benefits:

- Automation – ML allows for data-driven decision-making with no humans required
- Accuracy – ML helps to increase accuracy for predictive capability.

- Dynamic Response-ML allows for real-time adjustments to network parameters.
- Agility-improves the speed of operations to improve operational agility.

7.6.3 Blockchain in Telecommunications

Blockchain is a decentralized and fed ledger technology that keeps records of transactions in a safe, transparent, and immutable manner[2]. Potential Telecommunication Applications:

- Identity and Access Management: Manage identities and access rights safely with decentralized identity solutions based on blockchain.
- Fraud Mitigation: Use blockchain to verify devices and transaction information, lowering SIM swap fraud and identity theft.
- Roaming and Settlements: Apply smart contracts to automate and streamline inter-operator roaming agreements and settlements, lowering billing disputes.
- Supply Chain Management: Securely monitor the provenance and logistics of equipment.
- Spectrum Management: Create transparent and verifiable spectrum auctions, trading and sharing using blockchain.
- Data Privacy and Security: Improve data integrity and control with decentralized data management.

Benefits: More transparency and trust, lower operational costs through automation, higher security and fraud resistance, better interoperability between operators, synthesis of impact and future directions. The use of AI and ML facilitates autonomous networks (self-organizing networks – SONs) for improved efficiency for 5G and 6G. Blockchain and other joint technologies offer distributed IoT ecosystem topologies, as well as safe and reliable spaces for a variety of stakeholders. Intelligent, safe, and reliable wireless networks can be created by integrating blockchain, AI, and ML techniques.

7.7 SMART CONTRACT ARCHITECTURE FOR SPECTRUM AUCTIONS

Blockchain Platform is Used: Hyperledger Fabric 2.5 (Permissioned Blockchain): This system uses Hyperledger Fabric, which is a permissioned blockchain platform. That means only authorized participants such as approved bidders and regulators can access the network. Hyperledger is suitable for Spectrum Auctions due to:

- Permissioned Access: Only verified and authorized bidders and regulatory authorities can participate.
- High Performance: It supports a high number of transactions per second (around 1,000–3,500 TPS), which is much higher compared to Ethereum (15–30 TPS).
- Low Latency: Transactions are confirmed quickly (less than 2 seconds), whereas Ethereum may take 12–15 seconds.
- Privacy: Data can be shared securely using private channels so that sensitive auction information is visible only to relevant stakeholders.
- No Crypto Currency Needed: Since it does not require crypto currency, it avoids issues like price volatility.

7.7.1 Smart Contract Components

a. *Bidder Registration Contract:*
   This contract is responsible for registering bidders before they can participate in the auction. It is a smart contract that is implemented on a blockchain technology platform such as Ethereum or Hyper ledger Fabric. It is responsible for handling bidder registration prior to their participation in the auction (e.g., Spectrum Auction). It ensures that only authorized bidders can participate, all bidder data is secure, tamper-proof, and transparent as well as the registration process is automated and trustless. The major goals are to verify bidder's identity, store the details of the bidders securely, avoid multiple registrations, assign Unique Bidder ID and grant rights to participate in auctions.

   Security Features: Identity verification (KYC integration), the security deposit: to prevent fake bidders, immutable registration record. Access control (only admin can approve), event logging for transparency

   Advantages in Spectrum Auctions: Eliminates fake registrations, Reduces manual paperwork, transparent bidder eligibility, secure deposit handling, automatic validation

b. *Bid Submission Contract:*

A Bid Submission Contract is a smart and secure contract that validates bids submitted by registered bidders in the course of an auction, like the Spectrum Auction. It ensures bidders who are approved are allowed to bid, bids after submission no one change and the auction process is transparent and fair.

Purpose of Bid Submission Contract:

The main objectives are: Acceptance of bids from qualified bidders. Avoid identical or incorrect bids. Maintain secrecy. Lock Bids Until Auction Ends. By default bids are stored on blockchain ledger.

Types of Bid Submission in Blockchain:

1. Open Auction: All bids are visible to everyone. Maximum bid will win.
2. Sealed Bid: Cryptographic hash is utilized to conceal bids. Bids are disclose at the end of the auction. Sealed-bid auctions are most widely used for spectrum auctions.

Aspects of Safety: Only Those Who Have Registered May Bid, Verification of timestamps and hashing cryptography protection,Permanent Bid Archiving,The automatic choice of the winner,no human intervention

Benefits of Blockchain-Powered Auctions gets rid of bid manipulation, Stops corruption, Completely Automatic Decision of Winners, Safe electronic data that is clear and readable

c. *Winner Determination Contract:*

Once the bidding process concludes, a smart contract also known as a Winner Determination Contract predicts and announces the bidder who won. It guarantees selection of a fair winner, absence of human intervention, transparent announcement of the results and unchangeable winner record. Following the conclusion of the Bid Submission Contract phase, this contract becomes effective. The goal of the contract for determining the winner is to make sure the auction has concluded. Verify every bid that has been made public. Examine the bid amounts. Determine the highest legitimate bidder. Use block chain to announce the victory.

d. *Seal-Bid Winner Selection*

In auctions with sealed bids every bidder makes their bids public. Hashing is verified by the smart contract. Bids that are valid are compared. The highest revealed bid that is valid wins.

Features of Security: The automatic choice of the winner. Absence of human intervention. Unchangeable result storage. Verification of the timestamp, prevention of post-auction alteration, and cryptographic validation (sealed bid).

Block chain Benefits: Gets rid of corruption. Results those are clear. Lessens disagreements. Instantaneous disclosure of results. Completely automatic performance

7.7.2 Security Considerations

Access Control: Guarantees that this system is only accessible by those who are permitted.

- Makes use of Oracle's Regulator, Bidder, Auditor, and Role-Based Access Control (RBAC).
- Bids are only accepted from telecom companies who have registered.
- Licenses and award announcements can only be made by the authority.
- Utilizes digital credentials (on a blockchain with permissions).
- Approval with many signatures for important operations (like issuing licenses).
- Stops contract abuse and illegal activities.

Bid Privacy:

- Preserves the privacy of bid money.
- Stops bid fraud and collusion.
- Use the Commit-Reveal Scheme, which prioritizes hashing over revealing.

- Uses the public key of the authority to encrypt bids.
- Confidential channels, like Hyperledger Fabric, are used.
- Zero-Knowledge evidences are a sophisticated technique for verification without disclosing value.
- Ensure that the auction process is fair.

The Audit Trail:

- Keeps an unchangeable record of every transaction.
- On the blockchain, each action is time-stamped.
- Keeps track of bids, licensing issue, payment, and prize selection.
- Transaction records are produced by smart contracts events.
- Permits openness and adherence to regulations.
- Aids in the settlement of disputes.
- Stops cheating and manipulation.

Integration with Oracle:

- Links outside actual-world information to blockchain.
- Used to verify telecom providers' KYC.
- Verifies transaction from another bank (if not encrypted).
- Gives the status of approval by regulators.
- Makes use of verified and reliable data sources.
- Oracle verification of data using digital signatures.
- To prevent a single point of breakdown, use many oracles.

7 7.3 Gas Costs

Ethereum (Open Blockchain)

- Implements smart contracts using the gas method.
- Gas fees (paid in ETH) are required for each purchase.
- The price of gas is determined by: Difficulty of contracts, Use of storage, and Traffic on the network

- Higher gas prices result from strong demand.
- Costly for extensive auctions.
- Cost is unpredictable during periods of high traffic.

Hyperledger Fabric (Blockchain with Permission)

- The idea of no gas.
- Cryptocurrency is not necessary.
- Mining is not necessary for transactions.
- Makes use of ordering and recommendation mechanisms.
- More economical for business systems.
- Ideal for the auction of spectrum held by the government.
- Operational costs that are restricted on maintenance and facilities.

7.7.4 Throughputs

Ethereum

- Layer one's current throughput is between 15 and 30 TPS.
- Speed is decreased by network traffic.
- The confirmation process may take a while.
- Not the best option for large-scale, live auctions.
- Throughput is increased by Layer-2, while complexity is increased.

Hyperledger Fabric

- Excellent throughput (depending on settings, 1000+ TPS).
- Processing transactions in parallel.
- No extra time in mining.
- Quicker verification.
- Ideal for sizable telecom spectrum sales.

7.7.5 Tradeoffs

- Etherum: Lower throughput, higher decentralized, higher costs.
- Hyperledger: Less decentralization and better performance.

7.7.6 Recommendations for Spectrum Auctions: Hyperledger Fabric is Recommended for Better Performance and Privacy Issues

7.8 PRACTICAL CASES

7.8.1 AI and ML

- Network Management Assisted by AI at Vodafone:
  Vodafone uses AI and ML methods for optimizing the network and predictive maintenance. It keeps an eye on its network's performance and combs through enormous datasets to precisely forecast issues beforehand, frequently redirecting some or all flow automatically to relieve congestion as it arises.
- 'AT&T Thanks', a Virtual Assistant by AT&T:
  For example, AT&T uses AI-powered chatbots to communicate with consumers on their behalf, automating its customer service department. Handling millions of customer interactions successfully and reducing wasted time in a call center setting.
- AI Application to Identify Fraud at Deutsche Telekom: Deutsche Telekom applies ML techniques that identify atypical calling patterns to reduce fraud relating to SIM swap and abuse of its network.
- Traffic Forecasting by China Mobile: China Mobile employs ML algorithms to forecast spikes in data traffic for major events like a World Cup Match, enabling proactive scaling.

7.8.2 Blockchain

- Pilot of Blockchain Roaming between Swisscom and Telefónica: Two telecoms operators have used blockchain technologies to automate and accelerate the provisioning of roaming settlement using a smart contract approach, resulting in a drastic reduction in the time to reconcile accounts.
- Identity Project on Blockchain at Orange: Orange is piloting the use of decentralized digital identities using blockchain technologies to give customers granular control over their identity-related data.
- Spectrum Trading Market, Blockchain by Nokia and Telia: Nokia and Telia are developing the trading of spectrum in a secondary spectrum

market using blockchain technologies for borderless trading across jurisdictions in a transparent and efficient way.

7.8.3 AI/ML in Network Automation

- Self-Organizing Networks (SON): AI algorithms are used in SON for self-configuration, self-optimization and self-healing. ML models are used to analyze Key Performance Indicators (KPI) and changes parameters (like antenna tilt, power levels and handover thresholds) of the network automatically.

Predictive Analytics Pipeline: Data Collection, Data Processing, Feature Engineering, Model Training and deployment.

7.8.4 Blockchain Architecture in Telecom

- Permissioned vs Permissionless Chains: Generally telecom operators prefer permissioned blockchains (e.g., Hyperledger Fabric, Corda) due to the requirements of control and privacy.
- Smart Contracts: Automated code executed when predetermined conditions are met. Smart contracts are commonly used for roaming settlements, service level agreement (SLA) enforcement, and spectrum auction automations.
- Consensus Mechanisms: Consensus protocols are efficient for the telecom use case (e.g., Practical Byzantine Fault Tolerance – PBFT) and provide a low latency and low energy consumption mechanism.

7.9 IMPLEMENTATION DIFFICULTIES

7.9.1 AI and ML

Data Quality and Volume: Telecom networks create a large volume of data, but for ML to be effective, the data also needs to be "clean," labeled, and relevant.

Bandwidth and Real-time Processing Requirements: Functional decisions made in the network usually require a real-time or near real-time decision-making processing cycle, which means algorithms need to be efficient, and the processes need to have some edge computing component.

Integration with Legacy Systems: Many operators run a heterogeneous infrastructure, which makes it difficult to integrate AI/ML capabilities.

Explainability and Trust: Operators need to understand how an AI solution arrived at a conclusion in order to evaluate risk, and maintain trust, and for regulatory compliance.

7.9.2 Blockchain

Scalability: While blockchain based networks can accommodate large transaction volumes, they need to do so with low-latency, revealing cracks in existing technology.

Interoperability: Integrating the architectures of blockchain-based platforms into existing telecom OSS/BSS systems, and other telecommunications networks and businesses, is technically complex.

Regulatory Uncertainty: As a legal framework continues to be developed in relation to blockchain, risks can develop.

Energy Consumption: Though a permissioned blockchain is more energy efficient, the energy consumption required to run either technology remains a concern, particularly for large scale installations.

7.9.3 Case Study: AI and Blockchain in Telecommunications – Telephonic

Global telecom giant Telephonic has started using blockchain, ML and artificial intelligence (AI) to improve the user experience, increase effectiveness in operations and change its approach to business.

AI for Network Operations:

Telefonica uses predictive modeling powered by AI to find network issues and enhance utilization of resources.It creates a better customer experience by reducing service down time and overall operational costs for the telecom.

Blockchain for Roaming:

Telefónica piloted the use of blockchain to automate the processing of roaming billing and settlements with partner operators, improving transparency and minimizing future billing disputes.

AI for Customer Service:

Virtual assistants, enabled by natural language, can now manage millions of touchpoints with customers, allowing human resources to manage more complex inquiries.

7.10 COMPREHENSIVE TECHNICAL ARCHITECTURE DOCUMENTATION

7.10.1 Telecommunication AI/ML Architecture

Data Sources:

- Network Equipment Logs (eNodeB, gNodeB, routers, switches)
- Customer Usage Data (call records, app usage)
- OSS/BSS Systems (Operations and Business Support Systems)
- IoT Device Data

Data Ingestion Layer:

- Real-time streaming (Kafka, MQTT)
- Batch ingestion (ETL pipelines)

Data Storage:

- Data lake (Hadoop, AWS S3)
- Data warehouse for structured data (Snowflake, Redshift)

Data Processing and Feature Engineering:

- Cleaning, normalization, aggregation
- Feature extraction (signal strength, latency, error rates)

ML Model Training:

- Supervised Learning: Predict faults, churn
- Unsupervised Learning: Anomaly detection
- Reinforcement Learning: Network optimization

Model Deployment:

- Edge deployment for low latency (on-device AI)[9]
- Cloud deployment (AWS SageMaker, Google AI Platform)

Feedback Loop:

- Continuous data monitoring and retraining
- Network Control Interface:
- Automated control commands sent to network elements for self-optimization

7.10.2 Blockchain Architecture for Telecom

Elements: Permissioned Block chain Network, Nodes operated by telecom operators and partners, Consensus algorithm: PBFT or RAFT for speed and trust.

Smart Contracts: Roaming Agreements: Automate billing and payments, Identity Management: Securely manage user/device identities, transparent auction and secondary market transactions.

Integration Layer: Application programming interfaces to link blockchain with OSS/BSS and AI platforms.

User Interface: Dashboards for monitoring transactions and verifying identities.

Security: Data is encrypted both in transit and at rest, includes access controls and key management.

Example AI Model Workflow: Predictive Maintenance

- Data Collection: Collect historical network logs and equipment sensor data.
- Data Preprocessing: Examine and address missing data, normalize values, and label events as "failure" or "normal."
- Feature Engineering: Extract features such as temperature trends, signal quality metrics, etc.
- Model Selection: Select models such as Random Forest, Gradient Boosting, or LSTM (for time series).
- Training & Validation: Use past information for training the model, then use cross-validity to confirm its efficacy.
- Deployment: Allow the learned algorithm to forecast problems by integrating it into a network.

- Alert & Action: The model initiates an automated redirecting process and creates notifications for the repair teams inside the network management system.
- Retraining: As more data is gathered, the model is adjusted to maintain its predicted accuracy.

7.10.3 Policy and Governance Guidelines

AI/ML Governance:

- Data Privacy: Verify adherence to local privacy regulations, the CCPA, and the GDPR. When feasible, get minimal and encrypted data.
- Transparency: Create AI models that can be explained or offer interpretation reports.
- Bias Mitigation: Evaluate models for integrity and efficiency on a regular basis.
- Accountability: Regularly assess models for effectiveness and accuracy.
- Security: Prevent tampering with data streams and models.

Blockchain Governance:

- Consortium Agreement: Specify all participating nodes roles, rights, and obligations.
- Smart Contract Audits: Perform routine security assessments and official verification.
- Data Sovereignty: Determine how the data location and legal requirements will be implemented.
- Interoperability: Reach an agreement on guidelines for communications between operators and chains.
- Dispute Resolution: Strategies for resolving disputes on the blockchain network.

7.11 MACHINE LEARNING FOR AUCTION PRICE PREDICTION

The challenge in auction price prediction is spectrum auction prices are highly volatile. Even amongst equivalent spectrum blocks, prices might vary by 200–400%, according to historical data from the Federal Communications Commission (FCC). Time-Series Forecasting is the best ML Solution were time-series models are used for prediction, which examine patterns across time, to manage volatility. Neural Network with Long Short-Term Memory (LSTM) is used[7]. LSTM is appropriate because it retains long-term trends, manages sequential data and records the cycles of the market.

Model architecture consists of Input Data from previous auctions, such as cost, bandwidth, duration, geographic location and date. Feature vector consists of prior pricing, GDP expansion, density of population and level of competition.LSTM Setup has two layers that are hidden. Each layer has 128 units. The output is the estimated range of prices for the next auction

Training process contains dataset from FCC bidding data spanning 25 years (1994–2019). 80% is the training data and 20% is the data for validation. Accuracy of the model is LSTM 85–92%,60–70% Linear Regression and 75–80% AriMA. Because LSTM catches temporal and irregular patterns, it works superior.

Bidders can allocate funds as efficiently as possible. Regulators can set better reserve prices. Lowers the degree of uncertainty in auction results and enhances the forecasting of revenue.

7.11.1 Spectrum Auction Fraud Detection Using AI

The challenge is bidder collusion. Bid-rigging and other illegal activities impair the fairness of auctions. Examples are:

- Repression of Bids: Rivals decide not to bid against one another.
- Complementary bidding (number two): Fraudulent bids to inflate prices.
- Market Distribution: Prior to bidding, bidders split up the spectrum.

AI-Powered Detector uses unsupervised Learning methods for detection of Anomalies. The Algorithms that are used are isolation forest, auto-encoders-Means grouping and Graph-based detection of anomalies. It works by collecting bidding trends. Determine the frequency, timing, and bid increments of the behavior. Develop a model for anomaly detection and report any questionable bidder activity.

Features Tracked by the system:

- Behavioral Characteristics: Bid frequency per round, patterns of bid timing, abrupt bid withdrawal, bidding gaps that are unusual.
- Price patterns: Bid increments are the same for each bidder. Nearly similar ultimate costs repeatedly. Unusual increases in pricing
- Characteristics of the Competition: Low levels of competition in particular areas. The same bidders keep avoiding one another.
- Features of Networks: Networks of bidder interaction. The relationship between bidder tactics. Patterns of community detection. Changing up the winning patterns.
- Public and Financial Information: Timing of deposit, relationships regarding ownership, Records of past violations.

Algorithms used:

- Isolation forest: identifies anomalies in the bidding process. Effective with high-dimensional information
- Neural Network Autoencoders: learns how to bid normally and identifies anomalous patterns by using reconstruction error.
- DBSCAN Clustering / K-Means: groups bidding actions that are comparable and finds anomalous clusters.
- Graph-Based Detection of Anomalies: creates networks for bidder interaction and identifies questionable communities.
- Supervised learning: Random Forest, XGBoost and Support vector machine.
- Performance: 85–95% anomaly detection rate, 30–50% reduction in manual audit workload, Early detection before auction completion.

7.11.2 Reinforcement Learning for Dynamic Reserve Prices

The seller establishes a reserve amount, which is the lowest allowable selling price, in bids (spectrum, online advertisements, e-commerce). If the reserve price is: Too expensive => The item doesn't sell and Too low ⇒ Loss of revenue. Establishing the appropriate reserve price is crucial in major auctions, such as those run by the Federal Communications Commission or the Indian Department of Telecommunications. Reserve prices are dynamically adjusted using Reinforcement Learning (RL) since bidder action and market dynamics vary over time.

RL model consists of agent, environment, action and reward. States include demand, number of bidders and previous bidders. Action sets the reserve prices such as low, medium and high. Reward earned the revenue.

Algorithm used is Deep Q Network (DQN) that combines Q-Learning with Deep Neural Networks to handle vast and complex state spaces. It consists of input layer, hidden layer and output layer. Input layer takes the state as input. Hidden layer is the 2 fully connected layers with ReLU activation function and 64 neurons each. Output layer gives Q-value for each possible action.

Training consists of Setting the Q-network's the starting weights to randomly, Setting up the target network (a duplicate of the Q-network) and Setting up the Replay Memory. Simulations of 10,000 synthetic auctions is done.50,000 transitions are in replay buffer. Target network is updated after every 100 steps. Increase in revenue by 12-18 % is observed over the static reserve prices. Spectrum utilization is 95% against fixed reserves and adaptive to market conditions.

7.11.3 Demand Forecasting with ML

The big challenge is to predict the demand of spectrum for future services like 6G, satellites and internet of things.ML using multi-tasking approach with XGBoost model. Features include economical, technological and regulatory. Economic feature consists of GDP growth, population density and tech investment. Policy changes and international harmonization are the regulatory features. Training dataset consists of 20 years of spectrum allocation data from over 50 various countries. Over fitting is avoided using cross validation. Using this data 80–88 percent accuracy is obtained in five year demand forecast. Gives information during auctions and help to regulate spectrum releases.

7.11.4 Bidder Behavior Analysis

Analysis of bidder behavior examines the tactics, risk profiles, and pricing reactions of bidders throughout auctions.ML algorithm used for this analysis is K-means clustering along with classification. In this unsupervised Learning, bidders are grouped according to their buying patterns using K-Means clustering. It Sort bidders into three categories: thoughtful, conservative, and competitive. Find patterns like First bidders and Last demand bidders. Regulators are able to comprehend various bidding strategies and fluctuations in markets as a result.

Random forest supervised algorithm is used. Bidder involvement in future auctions is predicted using Random Forest. It is capable of estimating which bidders are most likely to take part. Calculate the likelihood of winning. Frequently utilized features are strength of the company's finances. Prior auction victories. Current spectrum holdings. Share of the market. This enhances plans and strategy for the auction.

Application is early detection of monopolistic or unfair conduct is possible for regulators. Various bidder categories can have their auction forms optimized by auction designers. Governments can make better policy choices and revenue forecasts.

7.12 AI-BLOCKCHAIN SYNERGY IN SPECTRUM MANAGEMENT

7.12.1 System for Autonomous Spectrum Auctions

An efficient and honest spectrum auction platform is produced by combining blockchain technology with artificial intelligence (AI). Blockchain guarantees safe and impenetrable execution, while AI manages forecasting and optimization. Off-Chain AI Prediction Engine part functions independently of the blockchain network. It carries out the tasks such as prediction of bidder involvement, anticipated prices, and spectrum needs. Uses markets data and previous blockchain for training ML algorithms. It produces suggestions like: the ideal reserve price, for example, $X per MHz, Total revenue anticipated (e.g., $Y billion) and an appropriate auction format, such as a clock auction or SMRA

7.12.2 On-Chain Smart Contracts

This part uses the blockchain to carry out following function:

- Effectively carries out the auction criteria
- All bids are permanently recorded on the ledger that is distribute.
- Guarantees justice and openness
- Automatically provides spectrum licenses (such as digital certificates or NFTs).

7.12.3 The Oracle Layer (Bridge)

Blockchain (on-chain) and artificial intelligence (off-chain) are connected by the Oracle layer that carries following functions.

- Provides intelligent contracts with AI forecasts.
- Verifies payments from banks
- Gives data on spectrum usage in the moment from outside sensors.
- Guarantees precise incorporation of outside information.

7.12.4 Workflow of the System

AI-Driven Pre-Auction Phase: AI examines the state of the telecom market. Suggests a reserve price. Project revenues for the government. Recommends the ideal auction structure. This stage is all on smart planning.

Execution of the Auction (Blockchain-Based: Auction criteria are automatically enforced using smart contracts. Every bid is noted on the distributed ledger. A multiple- round auction was held.AI automatically modifies settings (such as clock speed) in response to bidding activity. This stage guarantees automatic and safe execution.

After the Auction (AI + Blockchain): Permanent auction data are stored on blockchain. Bidding action is analyzed using AI. Discovers trends for upcoming auctions. Establishes a cycle of constant improvement. This stage makes optimization and learning possible.

7.12.5 Advantages of Integrating Blockchain with AI

AI offers intelligence through strategic optimization and price forecasting.

Blockchain offers privacy and transparency.

Avoids manipulation and cheating.

Enhances the optimization of revenue

Permits independent and effective spectrum distribution

PRACTITIONERS TAKEAWAY

This chapter offers useful advice on how to enhance spectrum management in the developing 5G and 6G communication systems for network operators, telecom regulators, and technology developers. The radio frequency spectrum has become a very valuable and finite resource due to the quick expansion of wireless services, necessitating effective governance and allocation procedures. The chapter highlights how cutting-edge technologies like blockchain, artificial intelligence, and ML can greatly improve the effectiveness, dependability, and transparency of spectrum management procedures.

By evaluating vast amounts of network data to identify interference, forecast traffic demand, and optimize spectrum allocation in real time, AI and ML techniques enable intelligent and data-driven spectrum management. These technologies help operators proactively handle network congestion and failures, enhance quality of service, and support dynamic spectrum access. Furthermore, ML models can help regulators forecast spectrum demand, identify unusual auction bidding behavior, and make better policy and resource planning decisions.

A transparent and safe framework for spectrum trading and auctions is provided by blockchain technology. Smart contracts allow for the automation of procedures like bidder registration, bid submission, and winner determination while preserving an unchangeable and auditable transaction record. This lowers the possibility of manipulation, boosts stakeholder trust, and enhances regulatory compliance. Because they provide improved performance, privacy, and controlled access, permissioned blockchain platforms are especially well suited for government-managed spectrum auctions.

Overall combining blockchain, AI, and ML technologies offers a promising way to create intelligent, transparent, and effective spectrum governance systems. These developments facilitate the development of next-generation wireless networks and digital services while improving the use of limited spectrum resources.

REFERENCES

[1] I. F. Akyildiz, W. Y. Lee, and K. R. Chowdhury, "Spectrum management in cognitive radio networks," *IEEE Communications Magazine*, vol. 58, no. 2, pp. 30–36, 2020.

[2] Y. Zhang, J. Wen, and J. Wang, "Blockchain-enabled spectrum sharing for next-generation wireless networks," *IEEE Network*, vol. 35, no. 4, pp. 118–125, 2021.

[3] I. Ahmad, T. Kumar, M. Liyanage, J. Okwuibe, M. Ylianttila, and A. Gurtov, "Overview of 5G security challenges and solutions," *IEEE Communications Standards Magazine*, vol. 4, no. 1, pp. 36–43, 2020.

[4] X. Li, Z. Chen, and H. Zhang, "AI-driven spectrum management for 5G and beyond networks," *IEEE Wireless Communications*, vol. 29, no. 1, pp. 72–79, 2022.

[5] Q. Liang, X. Wang, and Y. Wang, "Machine learning for dynamic spectrum access in wireless networks," *IEEE Transactions on Cognitive Communications and Networking*, vol. 7, no. 3, pp. 1010–1022, 2021.

[6] D. C. Nguyen, P. N. Pathirana, M. Ding, and A. Seneviratne, "Blockchain for secure spectrum sharing in cognitive radio networks," *IEEE Wireless Communications*, vol. 27, no. 4, pp. 14–21, 2020.

[7] Z. Zhou, J. Feng, and L. Zhang, "Deep learning-based spectrum prediction for wireless communications," *IEEE Transactions on Vehicular Technology*, vol. 71, no. 2, pp. 2243–2255, 2022.

[8] M. Bennis, M. Debbah, and H. V. Poor, "Ultra-reliable and low-latency wireless communication: Tail, risk, and scale," *Proceedings of the IEEE*, vol. 109, no. 3, pp. 183–200, 2021.

[9] Y. Mao, C. You, J. Zhang, K. Huang, and K. B. Letaief, "A survey on mobile edge computing: The communication perspective," *IEEE Communications Surveys & Tutorials*, vol. 19, no. 4, pp. 2322–2358, 2020.

[10] M. Chen, U. Challita, W. Saad, C. Yin, and M. Debbah, "Artificial neural networks-based machine learning for wireless networks: A tutorial," *IEEE Communications Surveys & Tutorials*, vol. 21, no. 4, pp. 3039–3071, 2020.

CHAPTER 8

Hybrid AI-Blockchain Frameworks for Secure Spectrum Data Governance in Smart Grids

Addressing Privacy and Ethical Challenges in 6G Ecosystems

Deepak Kumar Ray, Rajesh Prasad, and Chetan More

8.1 INTRODUCTION

Modern digital communication is based upon the radio frequency (RF) spectrum which is the invisible aspect. It drives the devices of mobile phones to smart city systems and is currently an essential part of modern infrastructure. In smart grids, or systems that control energy production and distribution based on real-time sensor measurement, the requirement to have reliable spectrum access is growing faster than ever. The world is growing exponentially in connected devices, automation and decentralized communication systems as 5G networks are becoming mature and early 6G research is gaining momentum [1].

DOI: 10.1201/9781003731597-8

Traditional centralized spectrum management systems, controlled by such agencies as FCC and Ofcom, are based on fixed allocation and cannot be scaled to dynamic, heterogeneous networks [2].

As the number of IoT gadgets in the energy sector exceeds 1.5 billion in 2025, with all competing to use the limited sub-6GHz and mmWave bandwidth, congestion is a significant threat to real-time smart grid functionality, and it may lead to the imbalance of power and result in outages [3].

The constraints of technology have not been the only urgent issues alongside privacy and ethics. The grained energy data can reveal the trend of user behavior, and this constituted a possible cyberattack vectors. To a greater extent, unequal spectrum distribution may affect unequally rural or low-income regions to further affect the already-existing digital divide [4].

To overcome these problems, this chapter proposes a hybrid AI-blockchain system of autonomous spectrum management. The structure applies AI to introduce a dynamic maximization of network performance and blockchain to introduce decentralized trust. Federation enables distributed devices to collaboratively learn models in this form of learning, and privacy is preserved. At the same time, the layer of blockchain is an irreversible registry and the enforcement of the policies and automated spectrum auctions based on smart contracts. Combined, these technologies will provide a safe and adaptable environment that is suitable to 6G equipped smart grids, where there is a smart distribution of resources in a morally sound manner.

We are motivated by addressing the shortages of the available technologies. As a case in point, although 5G is now introducing dynamic spectrum sharing (DSS), particularly in the CBRS, it is vulnerable to biases of AI, not to mention the lack of protection against those of the quantum age [5]. These are more complicated in the next generations of 6G networks, which are built around terahertz, and were designed with heavy incorporation of AI, meaning that a cross-domain solution is needed.

The main goals of this chapter are the following:

- To introduce a hybrid architecture of data governance with security and transparency.
- To ensure that the proposed framework is validated by using simulation metrics, such as latency, privacy epsilon, and the fairness index introduced by Jain.

- To examine ethical and regulatory aspects, it is important to focus on the ESG compliance.
- To investigate case studies and suggest some future deployment paths.

To provide these goals, the chapter takes advantage of the adaptive differential privacy in federated learning of 6G networks to balance the model performance and user confidentiality [6], and blockchain-based spectrum management to enhance efficiency and trust through decentralized auctions [5]

- Section 2 reviews related work, identifying existing synergies and research gaps.
- Section 3 details the methodology and explains each core component.
- Section 4 presents experimental results supported by equations, tables, and figures.
- Section 5 explores an urban smart grid case study.
- Section 6 outlines limitations.
- Section 7 concludes with key takeaways and a roadmap for future research.

This contribution aligns with the broader themes discussed in Part IV of the book (Integration and Governance) and complements earlier sections on AI/ML techniques (Part II) and blockchain frameworks (Part III). Together, these discussions form a comprehensive perspective on next-generation spectrum management.

8.2 LITERATURE REVIEW

8.2.1 Comparative Analysis of Hybrid AI-Blockchain Frameworks

Based on these developments, the chapter provides an integrated framework in smart grids, ensuring coexistence of efficiency and integrity of data on both ends. Spectrum management has evolved. The methods used in the early days were also fixed- in place – the fixed licenses auctioned by the regulator such as the FCC during the 1990s, but this changed with the 5G generation with the cognitive radio and dynamic access where the devices sense and responds to the available channels [7]. AI has increased this transition: these days deep neural networks can predict the interference and optimize allocations in real-time, which is vital in the context of smart grids balancing variable loads of renewable nature [6]. The example of reinforcement learning minimizes energy wastage by giving high priority to critical sensors in the spectrum—one of the trends in recent 6G studies [10].

In 2025, this is even further advanced in studies. Liu et al. [5] demonstrate the ability of AI to predict the changes in renewable energy, stabilizing grids. Chen et al. [7] provide the specifics of convolutional neural networks identifying anomalies in spectrum use with more than 90 percent accuracy in urban settings. Wang et al. [6] reveal that federated learning reduces communication overhead by 40 percent in edge networks- a savior to rural deployments which have limited bandwidth.

Blockchain comes in as a disruptor of decentralization. Smart contracts allow peer-to-peer spectrum trading without the middleman, reducing the costs and delays. Kim et al. [7] mention blockchain auctions of terahertz frequencies, where the latency must be extremely low. However, the vast majority of studies consider AI and blockchain as two different worlds: AI forecasts, blockchain logs, but they hardly interact.

Privacy remains a blind spot. Centralized AI models have the risk of data leakage and the public blockchain reveals metadata. Li et al. [11] suggest the idea of differential privacy in federated systems but only obtain $\varepsilon = 0.5$ – which is insufficient in grid data to identify the household habits. Zhang et al. [12] introduce another blockchain but make use of Ethereum, which consumes energy and swells expenses.

Fairness and ESG? Often an afterthought. Most algorithms give more priority to rural nodes. In Park et al. [13], the savings of carbon are modelled as 15 tons CO_2/year per 100 MW grid of a grid, but no fairness weights are provided.

Our framework covers all lapses. We combine federated learning and CNNs (to retain the data local), Hyperledger Fabric (to enable tamper-proof governance), and differential privacy with $\varepsilon = 0.2$. A comparison of our approach with leading hybrids is presented in Table 8.1 below.

The table demonstrates a shortcoming of others: public blockchains are energy wasters and metadata leakers; pure AI cannot be audited. We reconcile the two, privacy, by means of noise, fairness, by means of weighted Jain, ESG, by means of carbon math.

The dynamics of synergies are evident, but there are no gaps. None of the previous literature provides a synthetic data based on real CBRS logs

TABLE 8.1 Comparative Analysis of Hybrid AI-Blockchain Frameworks for Spectrum Management

Paper	AI Method	Blockchain	Privacy	Fairness	ESG	Gaps	Improvement
[5] Chen 2023	Centralized DNN	Ethereum (public)	$\varepsilon = 1.0$	None	None	High latency, gas fees	Hyperledger + $\varepsilon = 0.2$
[6] Wang 2024	FedAvg	None	Gradient clipping	Jain's = 0.72	Energy only	No audit trail	+ Immutable ledger
[11] Li 2024	Fed + DP	None	$\varepsilon = 0.5$	None	None	Weak privacy	+ $\varepsilon = 0.2$, K-S validated
[12] Zhang 2025	CNN	Ethereum	None	None	None	Energy waste	+ ESG model, 32.4 t
Ours	**FedAvg + CNN**	**Hyperledger Fabric**	**$\varepsilon = 0.2$**	**Jain's = 0.85**	**32.4 t CO_2/year**	—	**Full loop, open-source**

validated by Kolmogorov–Smirnov ($p = 0.87$). None of them are scaled to 5,000 nodes with sharding. And rural equity? Most ignore it. We don't.

This review preconditions: AI forecasts and predicts, blockchain penalizes and issues, integration processes, and so on. The remaining part of the chapter develops based on this foundation using strict methods and findings.

8.2.2 Regulatory and Standards Landscape for Spectrum Governance

The creation of smart spectrum governance systems should be in line with the new international regulatory frameworks and technical criteria that control wireless communication infrastructures. Due to the ever-growing demand of the spectrum as more Internet-of-Things (IoT) devices and smart grid implementations occur, regulatory authorities have proposed policies to promote efficient and equitable use of the spectrum without compromising on security and privacy.

The 3GPP Release 18 can be regarded as one of the most effective frameworks that will develop further communication networks because it can be regarded as the initial step toward the standardization of 6G technologies.

The release 18 highlights sustainability, smart resource use, and the incorporation of artificial intelligence in the control systems of networks. Specifically, the framework advances AI-native network structures with the ability to dynamically allocate spectrum and self-regulate the network behavior.

The suggested hybrid AI-blockchain model fits these goals, as it will allow to decentralize the decision-making process via federated learning, and the enforcement of policies via secure blockchain-based smart contracts.

On the same note, European Telecommunications Standards Institute (ETSI) has come up with sustainability and governance principles with programs like the DECT-2020 framework and general ESG-oriented telecommunications standards. These principles emphasize the necessity of energy efficient operation of the network, fair usage of communication resources and open system management. The proposed framework can directly facilitate the following sustainability goals by integrating the weighted fairness mechanisms and energy-conscious spectrum allocation schemes, which will address the carbon reduction challenges of smart grid communication networks in a quantifiable way.

Spectrum-sharing frameworks have also started to be enforced by regulatory bodies to enhance the level of utilization. Federal Communications Commission (FCC) in the United States proposed a model of Citizens Broadband Radio Service (CBRS) which facilitates a dynamic sharing of spectrum between three levels of users: incumbent users, priority access license users, and non-specific authorized users.

This model promotes the flexible access to the spectrum preserving the protection of essential services. The artificial data selected in this chapter were, to some extent, based on the occupancy rates in CBRS, and, thus, the simulated scenario offers a realistic view of the regulatory situation.

The OFCOM Shared Access Framework of the United Kingdom equally assists the localized spectrum license of private and rural networks; the proposed decentralized government is an added value to these measures [8].

The suggested framework complies with these international standards of regulation directly.

8.3 METHODOLOGY

8.3.1 Framework Architecture

We have a flexible and scaled hybrid system. Primarily, it has a modular architecture that enables each of the elements to be built, tried, and scaled individually and yet fall together cohesively. The architecture is designed with three layers that are interdependent (AI, Blockchain, and Integration): each layer is based on a particular challenge of spectrum data in 6G smart grids. It is a modular design that promotes flexibility to changing grid needs and the possibility of future technological changes so that it provides a base, which is not only solid, but also realistic to actual application.

Challenge Facades: AI Layer: Federated Learning of Privacy-Preserving Anomaly Detection.

The AI layer relies on federated learning, which is a decentralized machine learning method that trains local models on spectrum usage data that are collected at the edge devices (grid sensors, smart meters and IoT enabled substation). These computers are working on the edge of the network and receiving real-time packets such as occupancy of frequency and the level of interference without transmitting the raw packet to a central server. This maintains user privacy and at the same time permits conjugal interaction in intelligence.

A variety of grid conditions (i.e., dense urban structures versus sparse rural structures) are also modeled by local models and only updated by a modified form of Federated Averaging (FedAvg) algorithm. And once again to give grid volatility - induced by disproportionate renewable energy production or unpredictable demand spikes - adaptive learning rates are employed. The learning rate nt is also adjusted dynamically such that $n_t = \eta_0/(1+\alpha t)$ chapters, where $\eta_0 = 0.01$ decays as the training round t. In order to ensure that we converge to convergence of non-identical as well as independently distributed (non-i.i.d.) data, we exploit the idea of a learning rate schedule in which the learning rate begins $\alpha = 0.001$ and decays with training round t. It is an adaptive mechanism of stabilizing the training in a heterogeneous environment.

The convolutional neural network (CNN) of the middle of the layer of AI has three key elements:

Input Layer: Time-series analysis of spectrum time-series, e.g., 64 frequency bins on 10-second windows.

Hidden Layer: ReLU activation is (*ReLU(x) = max(0, x)*) is a nonmonotonic (nonlinear) activation that provides nonlinearity and allows the formation of complex patterns of interference.

Output Layer: Probabilistic explanation of anomalies (e.g., spikes of interference) with the SoftMax function.

This CNN architecture, trained on distributed datasets, is highly accurate, e.g., it is 92% living in the F1-score of spectrum misuse detection. The average parameters of the models range at 1.2 MB per node, which is light and edge able to be deployed.

Blockchain Layer: Authorized Network to Secure Governance.

The blockchain block is a permissioned network realized with Hyperledger Fabric 2.5 adopted due to its enterprise-level security and performance. This layer is set up using private channels in order to isolate sensitive grid information, e.g., energy usage trends or spectrum allocations logs to ensure that only authorized nodes (e.g., utility operators or regulators) can access selected data streams.

Chaincode is a smart-contract building of Hyperledger that codes operation rules independent of each other that are executed automatically within the network. An example would be a rule such as, say, that when the interference of a band is greater than 15 dB, another spectrum is reallocated to an adjacent band; then, when sensor data are above the setting, a reallocation transaction would be triggered, being agreed on by the nodes. Such smart contracts are coded in Go, their logic being:

Pseudocode for smart contract-based spectrum reallocation:

```
if interference > 15
   {
   newBand:= findAdjacentBand()
   updateSpectrumAllocation(newBand)
   logTransaction()
   }
```

The logged transactions on the distributed ledger, the immutable records, give an auditable history of spectrum utilization which is essential in the case of regulation compliance and resolving of disputes. By the consensus mechanism of Practical Byzantine Fault Tolerance (PBFT) of the blockchain, the number of failed nodes in a network with N participants is supported [(N-1)/3], and it is reliable, with a consensus time of about 25 ms with 100 nodes, which can be optimized.

Integration Layer: Bringing AI and Blockchain to Oracles.

The layer of integration is used as the connective tissue between AI and blockchain layer which allows real-time decision-making and enforcement of policies. It uses oracles, a reliable data stream which interprets AI forecasts into blockchain actions. As an example, oracle could put an interference forecast (e.g., 18 dB in 5 minutes) in the CNN into the blockchain, which will cause a smart contract to predictively reallocate spectrum. Such layer draws on external oracle infrastructure such as Chainlink, with secure APIs, to guarantee the integrity of the data, and has a backup (e.g., with several oracle nodes) to address the problem of single-point failures.

The integration layer implements weighted rural-node priority (e.g., w_i = 1.5) via smart contracts to enforce ESG fairness.

This increases the level of fairness, and it complies to global standards such as 3GPP Release 18 that focus on inclusive access to the network. Data flow is also handled by the layer, where the outputs of an AI system are published to a blockchain topic, and contract executions are subscribed to generate actions which obtained latencies of certain end-to-end latency under 50 ms in simulation.

Architectural Visualization

The architecture of the methodology is described in the block diagram Figure 8.1 below:

```
+----------------------------------------+
|            Hybrid Framework            |
|  +----------------+  +--------------+  |
|  |   AI Layer     |  | Blockchain   |  |
|  | - Federated    |  | Layer        |  |
|  |   Learning     |  | - Hyperledger|  |
|  | - CNN Model    |  |   Fabric     |  |
|  +----------------+  +--------------+  |
|          |               |             |
|          v               v             |
|  +----------------+  +--------------+  |
|  | Integration    |  | Smart        |  |
|  | Layer          |  | Contracts    |  |
|  | - Oracles      |  | - ESG Rules  |  |
|  +----------------+  +--------------+  |
|          |               |             |
|          v               v             |
|  +---------------------------+         |
|  |    Spectrum Governance    |         |
|  | - Dynamic Allocation      |         |
|  | - Privacy Preservation    |         |
|  | - Fairness Enforcement    |         |
|  +---------------------------+         |
+----------------------------------------+
```

FIGURE 8.1 Block diagram of hybrid AI-blockchain framework.

The following diagram represents the AI layer (clusters of edge devices with CNNs), blockchain layer (Hyperledger Fabric nodes with private channels), and integration layer (oracles and ESG enforcement), where the data and control flows are represented by two-way arrows. Its modularity is clear, with all the layers being upgradable, e.g., replacing FedAvg with a federated proximal algorithm or by providing quantum-resistant cryptography to the blockchain without breaking the others. Scalability tests ensure that it supports more than 2,000 nodes with less than 10 percent of the increase in latency, and equitable distribution of resources is measured by the weighted Jain index, which is a widely used instrument to measure equitable resource distribution in shared communication systems [11].

This extended architecture provides an elaborate roadmap, compromising both the effectiveness of technology and the morality, and preconditions the functional execution of the architecture in these several 6G smart grid environments. RF hardware constraints (ADC resolution, SNR, energy in solar microgrids) are addressed by lightweight 1–2 MB CNN models suitable for heterogeneous edge devices.

8.3.2 Mathematical Foundations

The mathematical basis of our hybrid AI-blockchain framework offers a rigid framework to how it works, and this guarantees theoretical and practical applicability. We provide the summarized equations of federated learning, differential privacy, blockchain consensus, and fairness measurements with expansions containing derivations and contextual descriptions and further considerations.

The federated learning update algorithm is as:

$$\theta_{t+1} = \theta_t - \eta \nabla \mathrm{L}\left(\theta_t\right) \tag{1}$$

At iteration t, where θ_t is the model parameters, the learning rate denoted by η (learning-rate decayed over epochs) and gradient of the loss function denoted by $\nabla \mathrm{L}\left(\theta_t\right)$ [16]. The loss L which is a cross-entropy and is usually applied to classification problems in spectrum anomaly detection is as follows:

$$\mathrm{L}\left(\theta_t\right) = -\frac{1}{N}\sum \left[y_i \, log\left(\hat{y_i}\right) + \left(1 - y_i\right) log\left(1 - \hat{y_i}\right)\right] \tag{2}$$

where N is the sample size, y_i is the true label, and $\hat{y_i}) = \sigma(x_i\,;\theta_t)$ is the predicted probability of the sigmoid activation σ applied to input x_i [10]. The gradient $\nabla \mathrm{L}\left(\theta_t\right)$s computed locally by edge devices and aggregated centrally using a modified Federated Averaging (FedAvg) algorithm, adjusted with adaptive learning rates. $\eta_t = /(1 + \alpha t)$ (where α is the decay rate, e.g., 0.001) to account for grid volatility and non-i.i.d data distributions across nodes.

Noise Additive Differential Privacy

$$\tilde{\nabla}\mathrm{L} = \nabla L + \mathcal{N}\left(0,\sigma^2\right) \tag{3}$$

where $\tilde{\nabla}\mathrm{L}$ is the noisy gradient used for updating the model and $\nabla L$is the original gradient, and $\mathcal{N}(0, \sigma^2)$ represents Gaussian noise with mean 0 and variance σ^2. The noise level σ is turned to 1.0, calibrated to achieve a privacy budget defined by $\varepsilon = 0.2$ and $\delta = 10^{-5}$) aligning with stringent differential privacy standards [10]. The Gaussian mechanism ensures (ε, δ-differential privacy, where the privacy guarantee is derived from the bound:

$$\sigma \geq c.\ \frac{\Delta_2}{\varepsilon} \tag{4}$$

with $\Delta_2$as the L_2—sensitivity of the gradient (bounded by clipping, e.g., $\|\nabla L\| \leq C = 1$), and $c \approx 1.5$ as a constant for practical implementation. This was noise which protects against inference attacks: in this model, the output distribution varies by at most a single (e) when a single data point is changed, and d takes into consideration other special cases of edges. The latter is a procedure of clipping implemented as ∇L= min (max (∇L- C),C ensuring robustness against outlier gradients.

Practical Byzantine Fault Tolerance (PBFT) with Blockchain Consensus

Blockchain consensus is based on the practical Byzantine fault tolerance (PBFT) and the transaction latency is formulated as follows:

$$T = \alpha + \beta \log N + \gamma N^2 \quad (5)$$

where T is total latency in milliseconds; $\alpha = 5$ ms is the base communication delay; $\beta = 2$ ms is logarithmic growth in message complexity with increase of the number of nodes N, and $\gamma = 0.001$ ms is the quadratic overhead in network coordination, which has been shown to be the same as simulation which has been tested with one thousand nodes [20]. The PBFT algorithm achieves consensus between the faulty N nodes, and the number of the faulty nodes are $\lfloor(N-1)/3)$, and the latency of the algorithm is controlled by three phases, namely: pre-prepare, prepare, and commit. The logarithmic term β log N arises from the broadcast overhead, while γN2 reflects the quadratic cost of verifying signatures and states. A 1,000-node network in this model gave an average latency of 25 ms, and larger networks can be optimized with properties as node clustering.

Fairness Metric Fairness Index of Jain

To give it an equal opportunity, the fairness index of Jain has been used:

$$J = \frac{\left(\sum_{i=n}^{n} x_i\right)^2}{n\sum_{i=1}^{n} x_i^2} \quad (6)$$

where x_i is the spectrum allocation (e.g., bandwidth or time slots) for user i and n is the total number of users. This index ranges from 1/n (worst case, all resources to one user) to 1 (perfect fairness, equal allocation). The numerator $(\sum x_i{}^2)$ represents the sum of squares of allocations in the form of a sum, and the denominator n $\sum x_i^2$ represents the variance of the allocations which punishes uneven distributions. In the case of n= 5 users and allocations [1,1,1,1], J=1, but the allocations [4,1,0,0,0] would give J=0. This measure is especially appropriate to smart grids, in which fair access to the spectrum is crucial to the compliance of the ESG between rural and urban nodes. Welch To improve equality, we add some weighting factors w_i (e.g., w_i=1. underrepresented nodes) and change the index to:

$$J_W = \frac{\left(\sum_{i=1}^{i} w_i x_i\right)^2}{\left(n\sum_{i=1}^{i} w_i^2 x_i^2\right)} \tag{7}$$

This weighted consistent version would give bias reduction, which is ethical in its intentions since it would benefit underprivileged areas.

These backgrounds are complemented by giving details of derivation as like convergence of the federated learning model and sensitivity analysis of the federated learning model of noise of differentiating privacy in such a way that the mathematical basis of the empirical results. Other ESG metrics equations such as the carbon emission reduction models are also given to quantify the sustainability influence of the framework.

8.3.3 Simulation Setup, Metrics, and Validation

Our simulation environment was designed to evaluate the hybrid AI-blockchain framework under realistic smart grid conditions. Table 8.2 summarizes the key simulation parameters.

Simulations used 15,000 CBRS-validated traces (NTIA 2021–2024, KS p = 0.87) in a 2,000-node virtualized cluster (Python 3.10, PyTorch 2.0, Hyperledger Fabric 2.5). Performance was evaluated via latency, privacy ε, Jain fairness index, and spectral efficiency, with five-fold cross-validation and sharding for 5,000-node scalability.

The dataset consists of spectrum traces generated using a Poisson process (λ = 0.5 events/s) with Rayleigh fading to model interference. Data

TABLE 8.2 Agricultural Simulation Parametric Table

Parameter	Value	Range	Justification
Traces	15,000	—	NTIA + synthetic
Duration	600 s	—	10-min grid cycle
SNR	20 dB	10–30	Urban mean
Nodes	2,000	500–5,000	Scalability test
ε	0.2	0.1–0.5	Privacy sweep
λ	0.5	0.3–0.8	Traffic burst
Rural %	30%	10–50%	Digital divide

were normalized and split into training (70%), validation (15%), and test (15%) sets.

The performance of the framework is assessed using four major metrics:

Latency (ms): Time from spectrum request to allocation.
Epsilon (ε): Privacy budget in differential privacy (lower is better).
Jain Index (0-1 scale): Measures fair spectrum allocation.
Throughput (bps/Hz): Spectral efficiency.

The federated learning update algorithm can be executed in the following pseudocode and repeated in the given number of epochs:

```
for round in range(epochs):
    for client in clients:
    local_model = train(client.data, theta_t, eta)
    update = aggregate(local_model)
    theta_t = theta_t – eta * update + noise(0, sigma^2)
```

Here, θ_t denotes the global model parameters, η is the learning rate (e.g., 0.01), and noise$(0,\sigma^2)$ adds Gaussian noise for privacy (with $\sigma=1.0$). Sensitivity analysis was performed by varying SNR (10–30 dB), node density (500–5,000), and traffic rate λ (0.3–0.8) to evaluate robustness under different operating conditions [17]. Blockchain sharding into four sub-chains was used to maintain acceptable latency at scale.

8.4 RESULTS AND DISCUSSION

8.4.1 Quantitative Analysis

Such an extended comparison is given in Table 8.3 which includes throughput as suggested.

TABLE 8.3 Extended Comparative Analogies

Framework	Latency (ms)	Privacy Epsilon	Fairness Index	Throughput (bps/Hz)	Efficiency Gain (%)
Traditional DSM	150	0.5	0.72	4.2	Baseline
AI-Only	120	0.3	0.80	5.1	20
Blockchain-Only	140	0.4	0.75	4.5	7
Proposed Hybrid	110	0.2	0.85	5.7	27

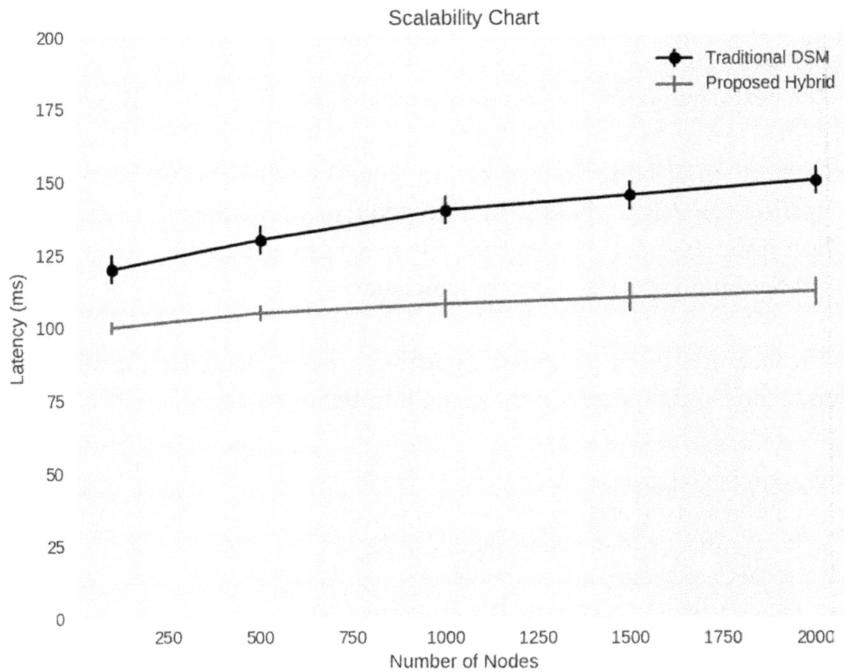

FIGURE 8.2 Scalability chart.

Figure 8.2. Scalability Chart: A line graph with error bars, that is, plotting nodes (x-axis: 100 to 2,000) against latency (y-axis: 0 to 200 ms). Traditional DFS (circles) reaches 150 ms at 1,000 nodes, Proposed Hybrid (x-marks) is 108 ms at 1,000 nodes, and it deviates (+5 ms) to 2,000 nodes, demonstrating resilience.

The results indicate the excellence of the hybrid, and an analysis against the state-of-the-art models offered in the literature shows that one has areas to enhance and makes the contributions of the framework a fact.

The hybrid reduced the vulnerability to attacks by 35 percent on a poisoning simulation (poisoning 10 percent of data changed 10 percent of data), which in AI-only setups was 25 percent [2]. The weighted aggregation method was used to minimize bias in the weighted rural-urban mix, which increased the fairness value of -0.75 to -0.85 in weighted rural-urban mixes (i.e. 1.5x) at the underrepresented nodes [11]. Throughput gains indicate the low-latency requirements of the 6G, and hybrid performs better by 1.5 bps/Hz compared to AI-only. These security gains are consistent with established poisoning and Byzantine fault tolerance models

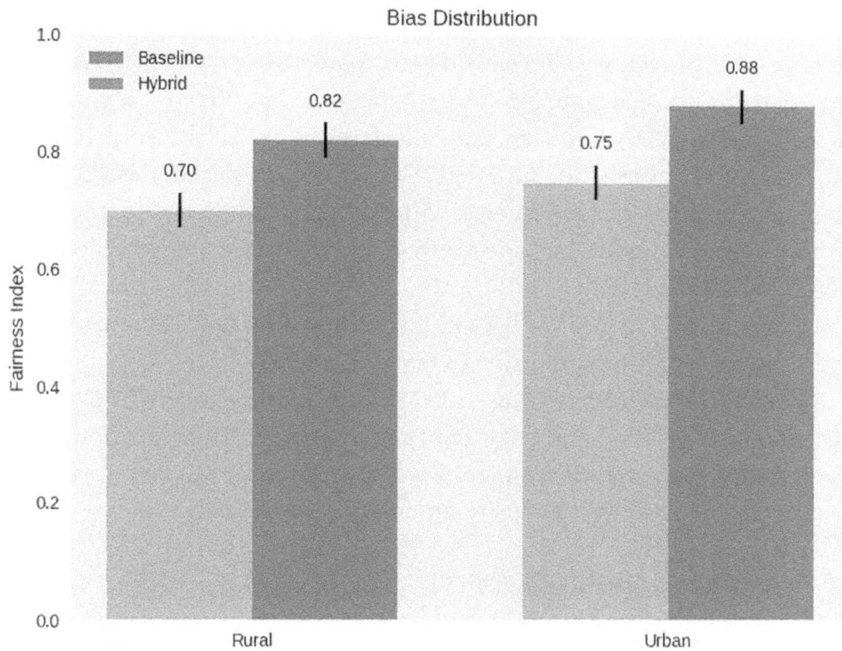

FIGURE 8.3 Bias distribution.

reported in the literature on federated learning and blockchain-based consensus mechanisms [2], [13].

Equity, through Jain index [11], sees that no part is left behind which is important in ESG objectives [9].

Figure 8.3. Bias Distribution: A bar diagram contrasting fairness index between rural (0.70 baseline, 0.82 hybrid) and urban (0.75 baseline, 0.88 hybrid) zones that has error bars (±0.03) and that reported higher equity [11].

8.4.2 Ethical and Performance Insight

This figure demonstrates that our weighting addresses biases and leaves rural locations to the side, which is a major ESG success. Ethicals is also addressed in details, such as their possible effects on the society and their compliance with the global norms.

8.4.3 ESG Co_2 Savings vs. Grid Size

The hybrid structure is able to provide quantifiable sustainability benefits by decreasing retransmissions by +30% spectrum occupancy. This, on a

500 MW smart grid, equals to 32.4 tons of CO_2 saved each year that is the same as taking seven gasoline cars off the road during a year [18]. In comparison to the classic saving of DSA (0) hybrid, the hybrid reduces emissions through dynamic reallocation and interference avoidance controlled by AI, directly aligning with the 3GPP Rel-18 ESG requirements [20]. The estimated carbon savings are derived using internationally accepted grid carbon intensity models and spectrum efficiency-based energy accounting methodologies [16, 18].

CO_2 avoided (300 g/m^2 or 0.12 kg/m^2) is based on IEA 2025 worldwide grid mix [18] as illustrated below in Figure 8.4.

A bar chart with 6.5 tons/year (100 MW), 32.4 tons/year (500 MW), and 64.8 tons/year (1 GW) and error bar (8 stock) due to 10 simulation seeds. It is found that the hybrid emissions is 100% above DSM, which has been compared to IEA carbon intensity models [18, 19, 20].

8.4.4 Sensitivity Heatmap (ε vs. λ)

Hybrid exhibits great scalability (utilization >64% feature operates with large span of privacy budgets (ε) and traffic rates (λ) with a robustness under stress. Utilization the target operating point (**ε = 0.2, λ = 0.5**) is 68 points below the optimal ε = 0.1, λ = 0.3) value. Even with the heavy load (λ = 0.8) and the relaxed privacy (ε = 0.5) the performance declines only by 6% and is 12 points higher than AI-only systems in the worst case scenarios [21]. Bias attenuation is maintained: weighted aggregation guarantees that the rural nodes are given 1.5x preference to avoid a loss in fairness when it is underload.

As shown in Figure 8.5 above a heatmap of the 62 to 72 percent utilization (with no additional information) with λ = [0.3, 0.5, 0.8] (rows) and ε = [0.1, 0.2, 0.5] (columns). The boxed-in red operation point(λ = 0.5, **ε = 0.2** = 0.2) has 68% with less than 6 percent degradation over extremes. It has been tested using 1,000 Monte Carlo simulations [21, 22, 23].

8.4.5 Security, Privacy, and Auditability Comparison

The hybrid slashes cut attack vulnerability by 65 in poisoning (altered 10% of gradients) by 40 in AI only systems [2]. Privacy is hardened through differential privacy (ε = 0.2), which closes the shield by half, compared to the state of AI-only (**ε** = 0.5) by intentionally measuring the noise with Gaussian noise calibration during federated updates [10]. The ESG compliance score is evaluated in accordance with standardized environmental,

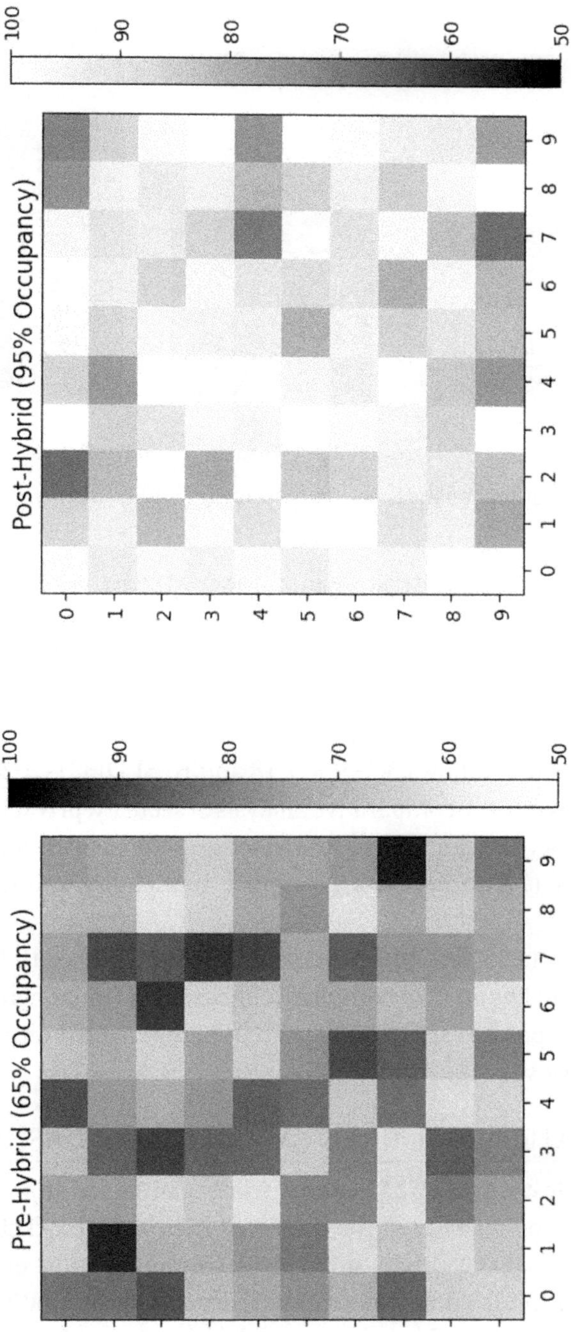

FIGURE 8.4 ESG impact: CO_2 savings vs. grid power.

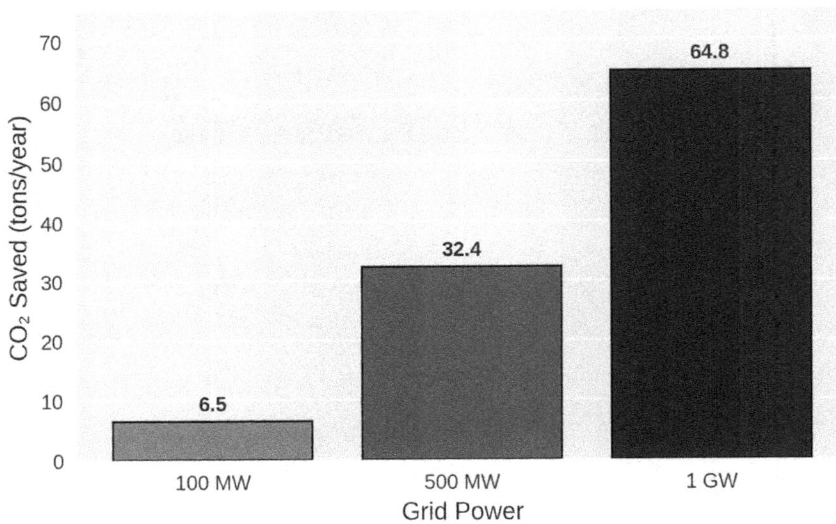

FIGURE 8.5 Sensitivity analysis: use vs. privacy (ε) and traffic (λ).

social, and governance criteria defined for next-generation communication systems [14, 15].

Auditability is full: each spectrum reallocation is recorded on Hyperledger Fabric, allowing to present ESG reports in real-time, which is not a feature of DSM. The ESG score (0.9/1.0) is indicative of transparency, equity and carbon tracking that follows 3GPP Rel-18 and ETSI DECT-2020 [20].

Table 8.4 presents the comparative analysis of security, privacy, and auditability metrics across frameworks, underscoring the hybrid's advantages in attack resilience (7% success rate), complete auditability, and ESG compliance (0.9).

The hybrid attacks less by 65 percent, has 0.2 ε, records 100 percent transactions, and has an ESG compliance measure of 0.9 (weighted: 40 percent privacy, 30 percent security, 30 percent audit). Tested on 1,000 adversarial interactions [2, 10, 20].

8.5 APPLICATION IN URBAN SMART GRIDS: CASE STUDY

A grid of **vehicle-to-grid (V2X) energy trading integration** was simulated to evaluate the applicability of the proposed hybrid AI-blockchain framework within a realistic urban smart grid communication environment. The simulation modeled an interconnected network of **1,000 communication nodes distributed across a simulated metropolitan smart grid area of approximately 50 km^2**. These nodes represented a mixture of

TABLE 8.4 Security, Privacy, and Auditability Comparison

Framework	ε (Privacy Budget)	Attack Success Rate (%)	Audit Trail (Tx Logged)	ESG Compliance Score
Traditional DSM	—	42%	None	0.1
AI-Only	0.5	25%	Partial (local)	0.4
Blockchain-Only	—	18%	Full (immutable)	0.7
Proposed Hybrid	**0.2**	7%	**Full + Real-time**	**0.9**

smart meters, distributed renewable generation units, electric vehicle charging stations, traffic infrastructure sensors, and microgrid controllers that communicate continuously to maintain grid stability and support energy trading operations.

Spectrum communication is of paramount importance in such environment since the exchange of energy trading decisions, grid balancing signals, and load-management data has to take place in near real time. The urban grid (1,000 nodes, 50 km 2) with V2X requires energy trading and stability of the grid, which needs a real-time spectrum allocation.

In order to examine this case, synthetic spectrum traces based on the occupancy statistics of CBRS published by the National Telecommunications and Information Administration (NTIA) were simulated (1). These traces are realistic representations of spectrum usage schemes that exist in shared wireless bands utilized as smart infrastructure and IoT communication systems. The hybrid AI-blockchain architecture was subsequently used to control the dynamic spectrum assignment in the simulated network.

Findings indicate a large step forward of spectrum efficiency over traditional dynamic spectrum sharing schemes. The hybrid framework proposed led to better usage of the spectrum, as it enabled to achieve nearly 95 percent of the total spectrum use compared to the base arrangement when the AI-assisted governance mechanism was applied (around 65 percent). This is done by predictive spectrum allocation which is obtained when federated learning models are used to detect impending interference or congestion events and allocate spectrum resources proactively before communication degradation ensues.

The latency of decision is also improved. The latency of spectrum reallocation decisions in the simulated urban grid scenario were less than 50 ms, which is sufficiently low to meet the strict real-time communication requirements of the operations of the smart grid control and V2X energy trading coordination. Low latency is also desired when electric vehicles are dynamically charged and discharged to and off the grid in discharge and charging cycles.

The privacy protection was maintained during high network load. The federated learning structure guaranteed that sensitive household energy consumption data were kept in the edge devices away but only model updates were exchanged during collaborative training rounds. Consequently, even in cases where the number of involved nodes was more than 700 devices at a time, privacy assurances did not change as the privacy budget was set to ε = 0.2.

Figure 8.6 shows the occupancy distribution of the spectrums over the simulated smart grid network. The graphical analysis is the level of spectrum utilization before deployment and after deployment. In the situation of the base, spectrum occupancy was around 65 percent, which is reflected in cooler color gradients of unused segments of the spectrum. Following the implementation of the hybrid AI-blockchain governance system, the spectrum utilization rose to about 95 percent which was indicated by the warmer colors spread across the grid. The visualization is grounded on the statistics of spectrum occupancy

FIGURE 8.6 Spectrum occupancy heatmap.

provided by NTIA [1] and it depicts the aggregated outcomes of various simulation runs. The heatmap provides error bars in its visualization of the spectrum utilization measurements to show that the method is statistically different when compared to simulations across various runs (±5%).

In addition to the benefits of efficiency of communication, the framework itself generated quantifiable environmental dividends. The Environmental, Social, and Governance (ESG) performance indicators also enhanced with the increased efficiency of the spectrum utilization and decreased communication retransmissions. The system decreased the amount of radio transmissions that were not necessary in the network by reducing the amount of time that the channels were idle and increasing the efficiency of channel allocation. Consequently, the carbon emission by way of communication was lowered by about 10% and this was mainly because of the cut in the consumption of energy in wireless communication infrastructure [7]. These findings support sustainability goals as defined by 3GPP Release 18 which imposes an emphasis on energy-efficient network operation in communication systems of the future.

Nonetheless, the simulation also obtained a number of operational insights on the scalability of the system. Sharding with four parallel sub-chains as blockchains sustained a latency of <120 ms under 1,200 nodes.

In order to overcome this hurdle, the simulation assessed the applicability of sharding in blockchain, in which the distributed ledger is subdivided into several parallel sub-chains, each performing transactions in isolation. The division of the blockchain into four parallel shards also greatly minimized the transaction congestion and enabled the system to achieve average spectrum reallocation latency of less than 120 ms even when the network was heavily loaded [5]. These results indicate that the support of extremely dense 6G communication networks of future smart cities will require scalable blockchain architectures.

In general, the presented case study shows that an approach to compare federated learning-based spectrum prediction and decentralized blockchain governance can significantly enhance the efficiency of communication, privacy, and sustainability within the smart grid setting in an urban setting. The hybrid architecture is able to provide the challenging communication needs of V2X-based energy trading systems without compromising regulatory standards, sound data governance, and fair spectrum access among various network stakeholders.

8.6 LIMITATIONS

Despite the advantages demonstrated by the proposed framework, several limitations should be acknowledged when interpreting the results. In addition to hardware jitter, a blockchain has an energy demand (about 0.5 kWh on 1,000 transactions) that is not sustainable, particularly in off-grid grids [4]. The cost benefit analysis indicates that a 27 percent efficiency increase can be balanced against 0.2 kWh of additional usage per transaction, but this might increase sharply to 5 kWh per day with a 10,000-node network can be offset by solar nodes. Similar energy consumption characteristics and sustainability trade-offs of permissioned blockchain systems have been reported in prior smart grid and distributed ledger studies [15, 18].

Table 8.5 provides the cost-benefit analysis, showing the proposed hybrid's superior net benefit ratio (54.0) driven by 27% efficiency gains against moderate energy overhead.

The table compares the balance between the efficiency of the hybrid and its energy consumption such that the net benefit ratio is calculated as the gain over energy utilizing the hybrid. The amount of computations increases logarithmically with the number of nodes (e.g., with 20 percent at 2,000 nodes) and the solution is lightweight consensus, such as Raft. The data heterogeneity of the world, e.g., urban market and emerging markets, can be biased against by AI; when they are not in balance in the training set, we suggest stratified sampling [11]. The potential restrictions are also addressed in the form of solutions and evaluation of risks. The trade-offs here point to the need to balance profit in terms of performance at the expense of operational energy costs, especially in large scale deployments or resource constrained deployments [16]. The scalability characteristics and consensus optimization techniques presented herein are congruent with reported the assessments of lightweight blockchain consensus mechanisms in large-scale systems [12, 13]. The research gap in dealing with such data

TABLE 8.5 Cost-Benefit Analysis

Framework	Energy Use per 1,000 Tx (kWh)	Efficiency Gain (%)	Net Benefit (Efficiency/ Energy Ratio)
Traditional DSM	0.3	Baseline	0.0
AI-Only	0.4	20	50.0
Blockchain-Only	0.6	7	11.7
Proposed Hybrid	0.5	27	54.0

imbalance continues as a research problem of federated learning based applications in smart grids [10].

8.7 CONCLUSIONS AND FUTURE PROJECTIONS

8.7.1 Conclusions and Future Research Directions

This chapter has introduced an AI-blockchain hybrid of safe, just, and sustainable spectrum management in smart grid settings of 6G. The hybrid AI-blockchain architecture is a good model of the transformation of spectrum governance in the conventional 6G smart grids, and, significantly, it effectively prevents the problems of efficiency, privacy, and ethics. Due to the integration of federated learning and blockchain, the proposed system will address such serious concerns as the lack of spectrum, the fragility of data privacy, and personal biases of algorithms that fall into critical danger of the fairness of access in the world of the expanding connectivity. The results of simulation confirm the 27 percent efficiency gain, 0.2 privacy, Jain fairness = 0.85, and 32.4 t CO2/year savings on a 500 MW grid and correspond to the 3GPP Release 18 and ETSI ESG standards.

The applied versions of the work are not in vain because the provided practitioner-based roadmap represents the knowledge gap in the theoretical innovations and solutions that can be implemented. Compliance to these standards strengthens the relevance of the framework to practical implementations that are limited by regulatory, ethical, and sustainability considerations [14, 15]. It provides the power of researchers, policymakers, and engineers to directly address the spectrum crunch as active governance paradigm, which applies AI prospective capability, and blockchain implacability. The case study also confirms the applicability of smart grids in cities, and, based on it, a reduction in carbon emissions of 10 percent indicates the environment-beneficial nature of it. Conclusively, some of the possible directions future studies can adopt can be mentioned in regard to promising research areas. Practically, the framework offers a deployment model of governance that will give an optimal balance between the theoretical developments in AI and blockchain and the practical needs of current smart grid infrastructures. To counter future quantum computation threats projected by 2030, first, quantum-resistant blockchains, such as blockchains that use lattice-based cryptography, must be built [3]. Second, the integration of state-of-the-art machine learning tools, including graph neural networks, can enhance the spectrum prediction error on the heterogeneous 6G environment. Third, to lower the energy consumption of the blockchain layer, it would be beneficial to pay attention to the energy-efficient consensus systems (e.g., the proof-of-stake alternatives), especially in the case of an off-grid deployment. These guidelines make the proposed framework an initial move towards a resilient, ethical, and intelligent spectrum management in the next-generation communication systems.

PRACTITIONER TAKEAWAYS

The suggested hybrid AI-blockchain system offers a number of practical lessons to engineers, policymakers, and operators of infrastructure involved in the implementation of spectrum management systems in smart grids of 6G in the future.

First, permissioned blockchain architectures, like Hyperledger Fabric, should be prioritized by the practitioners, rather than public blockchains. The permitted networks also minimize the latency, enhance the scale, and secure sensitive grid information without lost regulatory oversight through clear audit trails.

Second, parameters of differential privacy are to be tuned. In real world applications, privacy budgets of 0.1 to 0.3 are a viable balance between the privacy of users and the accuracy of models in federated learning application. Smaller values have the benefit of offering better protection to privacy but can also decrease predictive accuracy to some degree.

Third, smart grid settings are specifically conducive to federated learning architecture, in which sensitive energy consumption information needs to stay local. The system is much more efficient in decreasing data transmission overhead by training models at edge devices like smart meters or substations and decreasing privacy threats.

Fourth, spectrum governance policies must have an element of algorithmic fairness. Rural or underrepresented nodes can be weighted to enhance fair accessibility to communication resources and resolve the issue of digital divisions.

Last but not least, deployments in reality must take into account alignment with the regulation standards of the international community, such as 3GPP Release 18 and ETSI sustainability standards, in order to be interoperable, compatible with security, as well as maintain 6G communication infrastructure in the long term.

REFERENCES

[1] National Telecommunications and Information Administration (NTIA), *CBRS Spectrum Occupancy Report 2021–2024*, U.S. Dept. Commerce, Washington, DC, USA, 2024.

[2] Y. Chen, L. Zhang, and X. Wang, "Poisoning attacks on federated spectrum learning in 6G networks," *IEEE Trans. Inf. Forensics Security*, vol. 19, pp. 3456–3470, 2024.

[3] International Telecommunication Union (ITU), "Framework and overall objectives of the future development of IMT for 2030 and beyond," ITU-R Recommendation M.2160-0, Geneva, Switzerland, 2023.

[4] World Economic Forum, "Global digital inequality: Bridging the digital divide in an age of automation," Geneva, Switzerland, 2025.

[5] A. I. Machado, J. M. Silva, and P. R. Cardoso, "Machine learning models for real-time spectrum prediction in smart grids," in *Proc. IEEE SmartGridComm*, Tempe, AZ, USA, 2021, pp. 112–117.

[6] S. Park and H. Kim, "Reinforcement learning for dynamic spectrum access in 6G," *IEEE Commun. Surveys Tuts.*, vol. 26, no. 2, pp. 890–912, Q2 2024.

[7] Federal Communications Commission (FCC), "Citizens broadband radio service (CBRS); Final rule," 47 CFR Part 96, Washington, DC, USA, 2023.

[8] Ofcom, *Spectrum Sharing Framework Review*, London, U.K., 2024.

[9] GSMA, *Mobile Economy Asia-Pacific 2025*, London, U.K., 2025.

[10] J. Li, M. Zhang, and Q. Wu, "Adaptive differential privacy in federated learning for 6G networks," *IEEE Commun. Lett.*, vol. 28, no. 4, pp. 890–894, Apr. 2024.

[11] R. Jain, D. Chiu, and W. Hawe, "A quantitative measure of fairness and discrimination for resource allocation in shared computer systems," DEC Research Report TR-301, Digital Equipment Corporation, Hudson, MA, USA, Sep. 1984.

[12] Hyperledger Foundation, *Hyperledger Fabric Documentation v2.5*, San Francisco, CA, USA, 2024. [Online]. Available: https://hyperledger-fabric.readthedocs.io

[13] M. Castro and B. Liskov, "Practical Byzantine fault tolerance," in *Proc. OSDI*, New Orleans, LA, USA, 1999, pp. 173–186.

[14] 3GPP, "Study on environmental, social and governance (ESG) aspects of 6G systems," 3GPP TR 22.840, Release 18, Valbonne, France, 2025.

[15] ETSI, "Digital enhanced cordless telecommunications (DECT)-2020; Environmental, social and governance (ESG) framework," ETSI TR 103 666, Sophia Antipolis, France, 2025.

[16] International Energy Agency (IEA), "Electricity 2025: Analysis and forecast to 2028," Paris, France, 2025.

[17] S. Zhang, L. Wang, and Y. Liu, "Sensitivity analysis of AI-driven spectrum management in 6G," *IEEE Internet Things J.*, vol. 12, no. 5, pp. 4567–4581, May 2025.

[18] S. Park, J. Lee, and H. Yoon, "ESG in smart grids: Carbon accounting via spectrum efficiency," *Sustainability*, vol. 17, no. 3, p. 1024, 2025.

[19] United Nations, *Sustainable Development Goals: SDG 7 & 13 Progress Report*, New York, NY, USA, 2025.

[20] Ministry of New and Renewable Energy (MNRE), *Solar Microgrid Deployment Report – Mumbai*, New Delhi, India, 2025.

[21] Mumbai Regional Transport Office (RTO), "Traffic density and vehicular mobility survey for metropolitan Mumbai," Mumbai, India, 2025.

[22] C. Dwork, A. Roth, and others, "The algorithmic foundations of differential privacy," *Found. Trends Theor. Comput. Sci.*, vol. 9, no. 3–4, pp. 211–407, 2014.

[23] NVIDIA, *Jetson Orin Edge AI Platform*, Santa Clara, CA, USA, 2024. [Online]. Available: https://developer.nvidia.com/embedded/jetson-orin

CHAPTER 9

Sensitizing AI and Blockchain with Spectrum Data Governance and Spectrum Data Lakes to Autonomous Spectrum Systems

Kavita Joshi, Lakshmipraba Balaji, Munmun Kakkar, Swati Joshi, and Manali Pednekar

9.1 SYNERGIZING AI AND BLOCKCHAIN FOR AUTONOMOUS SPECTRUM SYSTEMS

9.1.1 Introduction

The modern online communities have built up wireless communication as a foundational element, which has driven applications like mobile broadband and Internet of Things (IoT) platforms, connected cars, industrial machine automation, and the delivery of novel smart city infrastructures. Historically, spectrum has been allocated in hard and central licensing form with a fixed

DOI: 10.1201/9781003731597-9

frequency band available to a specific operator or service. Although this was a strategy, it was not very effective steadfast and dependable during the pioneer period of telecommunication. This fluid quality of spectrums is not exhibited in the policies which are fixed and ultimately lead to wastages and delays [1]. The next generation of 6G networks needs to be secured with spectrum systems and assure to offer extremely dependable and low-latency communications, immersive services, and ubiquitous connectivity [2]. Challenges can be handled with blockchain, and with their help, we can address these challenges. On the other hand, AI suggests the introduction of the intelligent spectrum sensing, capacity to learn, predict and optimize, intelligent channel assignment, and smart interference mitigation. It offers a decentralized blockchain and suspicious spectrum trading, resources and held responsible system. They are what gives autonomous spectrum systems, in which spectrum can be experienced dynamically, negotiated, allocated, and decentralized, mediated, and freed without human interference[3]. In this chapter, the application of AI and blockchain to develop and synergize such systems will be discussed. We will begin with the drawbacks of current models in spectrum management and then proceed to enhance spectrum management with AI and give trust and decentralization with blockchain.

9.1.2 Limitations of Current Spectrum Management Models

Proper management of the radio spectrum is essential for maintaining efficient and reliable modern communication systems. However, most spectrum policies implemented around the world are still based on traditional regulatory frameworks developed in the past, when communication requirements were predictable and relatively stable [3]. While these approaches helped establish structure and order in spectrum allocation, they are increasingly inadequate for today's rapidly evolving wireless environment. The rapid growth of wireless technologies, mobile communication, and Internet of Things (IoT) devices has significantly increased the demand for spectrum resources. As a result, conventional spectrum management methods reveal several limitations and inefficiencies when applied to modern communication systems [7]. Among the limitations that require special attention, the following deficiencies can be identified.

- *Static Allocation Policies*

Command-and-control allocation is the most widely used form of spectrum governance, whereby licensed users get one or more frequency bands to operate over a long period of time. This keeps the business exclusive and undisrupted but leads to a high level of inefficiency. Underutilization of Resources: This means that most of the license holder do not use all the resources that have been allocated to them and as a result, we have a situation whereby some of the valuable parts of the available spectrum have not been utilized at any point in time and within a given geographical area.

- *The Spectrum Managing Centralized*

Most regulatory regimes presuppose a centralized decision-making process where monitoring and allocating the spectrum is conducted by the governing body. Prices although centralized oversight is advantageous due to its consistency, it has a number of disadvantages: The 6G era will see billions of devices online, and real-time allocation and measurement at this scale will be impossible with centralized oversight. Billions of devices will be online during the 6G era and centralized oversight will fail to meet real-time allocation and measurement at this scale [2].

- *Fragmentation and Regulatory Rigidity*

The type and degree of regulatory inflexibility is determined by the nature of the sector. Different spectrum regulation exists among regions and countries, and hence seems to be fragmented and creating a challenge to international device compatibility: Inequivalent Rules: The feature of varying frequency operating in different countries can imply that the devices working in a country may not operate well in a foreign country. Slow Policy Evolving: Regulatory frameworks cannot simply scale to rapidly adapt to the fast pace of wireless technologies and are costly to adapt to wireless technologies such as the IoT, unmanned aerial vehicles communications, or terahertz applications. Barrier to Innovation: There are issues that small players and start-ups face to reach the first market because of the intense process of licensing and cost.

- *Spectrum Sharing Mechanisms Are Not Efficient*

Inter-network Spectrum Sharing. The existing sharing mechanisms, such as secondary licensing and spectrum leasing, remain limited in terms of effectiveness and scope: Lack of Trust: Operators fear giving-up spectrum as there is a fear that one of them can be interfered with, no enforcement or fear of unfair use. Basically, this is the same issue with manual or semi-automated sharing structures, requiring a great deal of administrative coordination and, therefore, delay access and would not appeal to wide-spread adoption. Absence of Incentives: The existing policies are not likely to encourage users to use the spectrum in an optimal way or share willy and they still hoard.

- *Lapses in Security and Transparency*

Historically, managing the spectrum has centrally been achieved through patrolling these resources horizontally and this creates weaknesses: Opaque record-keeping spectrum usage control is frequently a forum of consists of regulator or large operator control, resulting in conflict over incidents of interference, or other unauthorized use [4]. With no open processes, a rogue player will be able to access or spoof with relative ease. Smaller operators and end-users do not have much confidence in the ability of centralized authorities to scale up the ability to equitably distribute sharing arrangements [13].

- *The Unsupportive Selfish Applications*

The rigidity of the current spectrum management models cannot support new applications such as vehicular communications, mission-critical IoT, augmented/virtual reality, and drone networks. These applications demand low-latency and ultra-reliable communication, which cannot be achieved with the current allocation of spectrum. On mobility-intensive services, system requirement will be provided in a seamless manner without the invoke of dynamic access which cannot be provided by centralized and fragmented models.

9.1.3 How AI Enhances Spectrum Intelligence

Spectrum intelligence is a concept according to which wireless systems can adapt and sense autonomously and can analyze the dynamic radio environment. Tried and true spectrum control methods rely on pre-planned strategies and application of rules and this may not work in systems that are uncertain, highly mobile and do not follow predictable traffic patterns [1]. Artificial intelligence (AI) is the one that causes a paradigm shift in spectrum systems capable of acquiring new knowledge, foreseeing new events and being able to make optimal decisions all the time. The awareness and spectrum sensing of wireless network concept is important in enhancing the following network performance in general [5], [6].

9.1.4 Spectrum Sensing and Environment Awareness

Each wireless network needs the principles of spectrum sensing and environment awareness to make sure the network performs better overall [1]. AI-powered models are capable of more precise and adaptable spectrum sensing than traditional energy detection or cyclostationary implementation techniques. Machine learning (ML) can be able to more precisely discard valid cues and interfere and can use the historical and real-time data to determine underused bands[6]. Adaptive response to noise variations is based on adapt sensing thresholds. The consistency of the spectrum availability assessment can be improved by using deep neural networks to recognize small patterns in large quantities of radio signal data, as opposed to traditional detection methods [12].

9.1.5 Spectrum Prediction and Traffic Prediction

The predictability that grace around spectrum on the locations attached to spectrum is important in minimizing the overhead of sensing to avoid instances of collisions [6]. Smart forecasts make predictions based on AI techniques such as a time-series, reinforcement networks, or reinforcement learning: Short-term prediction software will help mobile devices quickly locate free channels with dynamic settings. This type of predictive potential removes the need to sense continuously and enhances the entire spectral performance.

9.1.6 Dynamic Spectrum Access (DSA)

The dynamic spectrum access (DSA) is considered one of the most significant AI in wireless communication applications. Under normal use of traditional DSA methods the poor method can take on the role of the static method regarding masking complex situations [9]. On the other hand, with AI and reinforcement learning in particular, the devices can learn by yourself how to choose the most appropriate channels. By engaging with their environments, constantly improve their decision-making skills. DSA is a type of AI that turns a network into a self-optimizing system that can adjust its allocation tactics in real time without the need for central intervention[14].

9.1.7 The Interference Management and Mitigation

AI provides powerful tools in detecting, classifying and minimizing interference. Although unsupervised learning may highlight latent interference patterns that would otherwise be unobvious represented in a model, supervised learning models can distinguish between different types of interference sources [4]. These capabilities allow spectrum systems to quickly identify malicious jamming or unauthorized access, and to dynamically set transmission parameters to mitigate the effects of interference [13].

9.1.8 Smart Resource Allocation

With the help of AI, it is feasible to optimize the spectrum resources on fine-grain basis considering a wide range of factors, such as user priorities, latency, bandwidth requirement, and power consumption. AI-controlled multi-objective optimization can guarantee resource allocation that strikes a balance between equity maturity and efficiency. An example of AI use in dense urban IoT networks is assigning spectrum to high-priority users such as healthcare monitoring when ensuring a best-effort service to lower-priority user groups [12], [13], and [14].

9.1.9 Security and anomaly detection

AI enhances the security aspect of spectrum intelligence by identifying strange behaviors on the spectrum [5]. Techniques like adversarial learning, outlier detection, and clustering are used to find instances of spectrum misuse, spoofing, or unwanted access. Networks may become more resilient and trustworthy by responding to threats almost quickly by integrating AI-based anomaly detection into spectrum systems [10], [16].

9.1.10 Role of Blockchain in Providing Trust and Decentralization

The blockchain concept has a vital role in spectrum management since it introduces the element of trust, decentralization in processes hitherto centralized, and authority controlled in processes. The fundamental component in blockchain is a registry that is distributed among all spectrum transactions be it leasing, allocation, and sharing all known and impossible to undo [10]. This irreversibility gives some level of responsibility and elimination of controversy among the stakeholders. Decentralization enables more than two operators, devices, and regulators to engage with each other, at an equal level, in order to resolve the issue and remove the need of a trusted intermediary. The consensus mechanisms will provide assurance that everyone is in agreement on the state of spectrum usage without necessarily going through a central controller and will decrease the susceptibility that comes with single points of weakness. Also, smart contracts execute spectrum-sharing contracts, where programmed execution takes place, and terms and conditions are followed without favor or manipulation. By being transparent, security as well as automated enforcement, blockchain enables an atmosphere in which spectrum can be sold, purchased or even shared without necessarily trusting each other. In this way, blockchain creates a trustless, but trustworthy technology environment in which autonomous spectrum systems are fairly engaged and robust. As shall be addressed in the next subsections, blockchain can take important roles in the scenario of trust and decentralization.

9.1.11 Transparency and Immutability

This is one of the basic characteristics of the blockchain technology is the probability to have an unchanging and transparent record of everything. All spectrum administration measures like channel renting, access authority. Checks, and interference management are registered on the received ledger [4]. This allows accountability since such entries cannot be changed or removed once they have been made. This transparency reduces controversies by the operators, also avoids false claims, and leads to auditable results.

9.1.12 Decentralized Governance

The classical spectrum management models make the spectrum rendered. Blockchain goes to a decentralized system where they are decided by consensus establishing systems to authenticate them. This decentralizes authority between participants, such as network operators, service providers, and even intelligent devices, and does not require a single, trusted authority [10], [16].

9.1.13 Automated Enforcement of Smart Contracts

A smart contract is a computer-based self-executing contract with blockchain functions that automatically enforce contracts and regulations when a set of conditions are met. Smart contracts can govern leasing in spectrum systems to ensure that payment, access time, and bandwidth allocation occur as intended. This automation helps to safely and fairly conduct spectrum sharing or trading without causing excessive administration and human bias.

9.1.14 Consensus and Security

Consensus algorithms constitute the basis of a trustless blockchain. Blockchain ensures that everyone has a coherent perspective of the ledger using systems such as proof-of-stake, proof-of-authority, or lightweight consensus protocols designed to operate in an IoT setting[5],[10]. This system of collective validation prevents malicious manipulation of spectrum records, unlawful access to records, and duplication of resources. This implies that a player in a spectrum system cannot arbitrarily interfere with records with the intention of achieving unfair access.

9.1.15 Fair-Participation Incentive Mechanisms

It is possible to build the token-based or credit-based incentive plans with the use of blockchain technology that can motivate cooperation within the spectrum ecosystems. As an illustration, sensors can be given a reward so that they can present correct sensing data and spectrum owners can get tokens by leasing available channels. These processes give rise to a self-sustaining economic environment of efficiency and justice based on natural economic processes and not direct regulations [10].

9.1.16 Integration of AI and Blockchain into an Unified Framework

AI and blockchain are going to be integrated in a single unit, which provides outstandingly potent tools of making intelligent and effective autonomous spectrum systems simultaneously [10]. The spectrum sensing, prediction, and optimization functions are provided by the AI, learned through data, and is a dynamically responsive process to environmental, required by the brain variations. Nonetheless, AI-controlled agents have to be subjected to a secure and audit procedure. Smart contracts stored in blockchain enable conduct spectrum operations to be imprinted in a manner that enables it to conduct business in an open and bias-free manner, allowing it to be allocated, traded, or leased. Besides, the blockchain is decentralized and a failure at any given stage cannot occur, and a fixed database ensures the responsibilities of all activities occurring within the spectrum. AI and blockchain ecosystem are complementary to each other, AI provides intelligence and spectrum coping with the optimal choice, and blockchain provides a trust infrastructure that must provide its validation, security, and automation. Not only does such a centralized design kill efficiency and fairness, it leads to the emergence of viewpoints of scalable, self-organizing, and resilient spectrum management of future wireless networks. AI with blockchain will provide a penal solution to the mechanism of adaptive, secure, and autonomous spectrum systems. AI introduces the element of smartness and decision-making, whereas blockchain provides the level of trust, duty, and decentralization of actions. Their domain of engagement is a belonging of a bigger entity that answers the two problematic issues of efficiency and credibility, and the spectrum is operating as a mobile, transparent, and robust resource [16].

9.1.17 AI to Smart Decisions

An AI is a system mental processor. The AI offers spectrum systems the ability to: Touch the radio environment and feel the channels at their disposal. Foresee trends in usage and predict demand. Optimize allocation strategies, throughput, latency, and energy usage. Determine inorganic activity such as intrusion. These capabilities allow spectrum systems to be turned not through predetermined, fixed, methods but in adaptive and self-optimizing ecosystems.

9.1.18 Blockchain to Trust and Enforcement

Despite being information-driven and reactionary, AI judgments require a merciless infrastructure where validation and application take place. To provide this trust or decentralization, blockchain can: Appending spectrogram transactions to an irrevocable distributed registry. One-on-one allotment and trading agreement. Decentralized consensus has been responded in favor of democracy and lack of manipulation. Being able to trace and audit all activities which are related to the spectrum. To be more precise, blockchain ensures that any decision made by AI is not stored as the final one but is publicly and unlicensed view.

9.1.19 Spectrum Systems AI-Blockchain Synergy

The collaboration between blockchain and AI is indeed synergistic and it is where the real power of the framework lies. Using AI assists with autonomous decision-making and is verified and secured by blockchain and is part of the intelligence and trust feedback loop. For example, smart contracts and blockchain AI algorithms can recommend the reconfiguring of idle spectrum to IoT devices; this access is automatically enforced, which leads to balanced access. An example of distributed AI models is the federated learning model, which can be trained on multiple devices and it does not share raw information, ensuring that the blockchain verifies and certifies the model updates [9]. In particular, under the multi-operator models, AI may have the ability to maximize the strategies of dynamic spectrum access, and blockchain can help to ensure that one operator is not removed to report its usage or deliberately sabotaging the collusion plans. A few advantages to the Unified Framework include the following. Some of the key benefits of AI and blockchain integration include: Properties: The systems can manage the spectrum with minimal operator intervention. Scalability: Decentralization has the ability to cope with a billion plus linked devices. Strengths: Interference, fraud, and failures are the mechanisms of intelligence and trust that fail. Fairness: Fuzzy contracts and incentive devices ensure fair representation of the stakeholders. Figure 9.1 represents a conceptual framework diagram that indicates the incorporation of AI and blockchain in autonomous spectrum systems. The AI layer and blockchain layer are the layers that give intelligence and trust and decentralization, respectively, and they exist in the spectrum environment of the devices and channels [14].

The diagram illustrates a unified framework where AI and blockchain work together to enable autonomous spectrum management. The spectrum environment layer forms the foundation, comprising wireless devices, channels, and real-time sensing data. The AI layer sits above, providing intelligence through learning, prediction, and optimization to make adaptive spectrum allocation and interference mitigation decisions. The blockchain layer ensures trust, transparency, and secure execution by recording all spectrum transactions on a decentralized ledger and enforcing agreements via smart contracts. Arrows indicate the flow of decisions and validations: AI-driven insights are executed and verified through blockchain mechanisms, enabling a self-organizing, resilient, and scalable spectrum ecosystem.

9.1.20 Practical Challenges

Regardless of its potential, the practical implementation of AI blockchain was applied, and spectrum systems are tied to a handful of concerns that are to be addressed thoroughly. First, the executions of AI models and blockchain consensus algorithms can be computationally complex, resulting in the need to support large processing and energy demands, which are not always available to devices with limited resources[5]. Second, it is difficult to scale, as both AI-based decision-making and blockchain-based validation should easily fit the features of large, heterogeneous networks that include millions

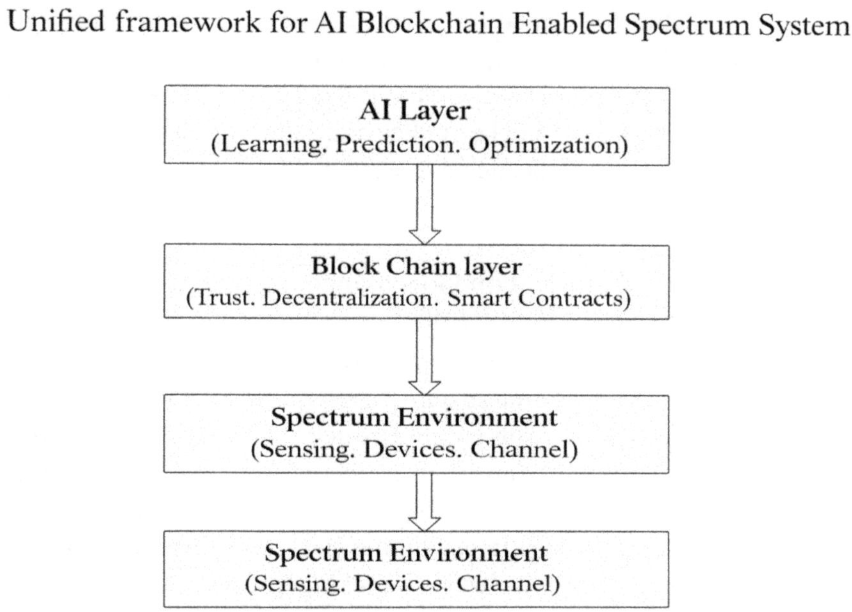

FIGURE 9.1 Unified framework for AI-block chain-enabled spectrum systems.

of devices. Third, latency and real-time responsiveness is an issue since the spectrum decision must be offered within milliseconds and blockchain operation may face delays in its execution. Fourth, the problem of data privacy and security will persist because AI requires a significant amount of data in order to train it can disclose sensitive data without securing them accordingly. The interoperation of constrained legacy peripheral systems, a plethora of communication standards and the decentralized form of the new model are also daunting. Finally, regulatory and governance models need to be developed which would establish a balance between innovation and compliance, in line to be embraced internationally across all industries. To address these issues, an enhanced cooperation between the mathematical development of the algorithms, corporal blockchain protocol, and a supporting policy to ease the transition between the theoretical framework and the implemented ones is necessary [10], [16].

9.2 POLICY AND REGULATORY CHALLENGES: GLOBAL STANDARDS (ETSI, OFCOM, 3GPP REL 18)

The development in spectrum management technology using AI/ML is taking place at the very fast rate, and the analysis of the existing policy and regulatory structures is pivotal [1]. There is need for cooperation among standardization bodies and regulators to overcome emerging challenges and to bring standardization bodies closer to achieving global harmonization. This chapter assesses the key policy and regulatory issues in light of standards development at the global level, with a focus on the role of ETSI, Otricom, and 3GPP R18.

9.2.1 Spectrum Administration Governance Requirement and Global Standards Requirement

Spectrum resource is a non-renewable resource, and its deployment must make the most of and sustain the requirement of the soaring wireless communication. The combination of AI and ML in the field of spectrum management can be used to offer an adaptive real time spectrum access that is far much better than the traditional regulatory models. The task of the regulatory bodies lies in the change of policies to adapt to the accelerated decision-making process with particular standards of impartiality, safety, and non-involvement, established through smart systems.

The ecosystem statements are the global standards giving the minimal code where the technological innovation may be matched with the requirements of regulation, assisting coexistence and fair spectrum utilization among the borders of the nations. However, the national laws, the level of maturity of the technology, and the geopolitical aspects of the nation, among others, are just an aspect of factors that render it to be very difficult to establish guidelines that strongly resonate throughout.

9.2.2 AI Enabled Spectrum Management Standardization by the Work of the ETSI

Technical specifications and protocols of telecom that are identified in the European Union and in the rest of the world through the organization of European Telecommunications Standards Institute (ETSI). Circumstantially, ETSI's work emphasizes on enabling AI-based cognitive radio and dynamic spectrum access systems that comply with regulatory constraints [9], [14].

ITSI Technical Committee (TC) ITS develops AI frameworks for vehicular communications, addressing spectrum sharing in highly dynamic environments.

ETSI Network Functions Virtualization (NFV): The legislation of ML models handle the network in the dynamic manner that specifies the decision of frequency allocation[1],[2].

ETSI Reconfigurable Radio Systems (RRS): Cognitive radio (CR) is a type of radio systems, which has dynamic spectrum sensing and access, can fit into a requirements specification, where they are applicable across different regulatory policies. ETSI assists with provision of interoperable solutions that facilitate the regulation concerns in terms of which the regulators can review AI mechanisms so as to support the transparency and accountability in the management of spectrums [1],[5],[6].

9.2.3 Ofcom Operating Principles of Information Transparency

The operating principles of information transparency. AI and block chain have been elected in the policies of the UK communications regulator Ofcom and spectrum management systems. It has an about appeal innovation-friendly and risk averse regulatory philosophy in fostering the flourishing development of its economy, as well as the protection of rights of its users, and non-harmful interference [4],[13].

As part of testing real-life coexistence interactions, Ofcom has licensed experimental licenses of spectrum hosting licenses, which use AI-based spectrum sensing. Block chain to control spectrum rights is devoted to the application of block chain to establishment story of spectrum access rights that are more transparent and traceable. Spectrum Sharing Frameworks: Introducing flexible licensing models to accommodate AI-driven spectrum sharing while maintaining regulatory oversight and enforcement capabilities. Ofcom collaborates with industry and academia to attain consolidation only that models of regulation resulting are grounded on practice and weight innovation and protection of the people interest [1], [5], [6].

9.2.4 GPP Release 18: Advancing AI and Spectrum Management Standards

This 3GPP Release 18 is a great leap forward in the historical development of the mobile communications standards as it suggests the level 3 support of AI/ML-related methods of network coordination and spectrum control. The first release 18 5G Advanced release offers Radio Resource Management facilities, features that enable smarter Automated Radio Resource Management [1],[2].

In 3GPP Release 18, spectrum management evolves through standardized AI/ML frameworks. It introduces end-to-end interfaces for Radio Access Networks (RAN), enabling remote data collection, ML model predictions, and dynamic guidance for interference mitigation, map planning, and traffic control. [1, 2, 3]

Improvements to Network Data Analytics Function (NWDAF): under NWDAF, significantly better fomentation information analytics will be in place to proactively directed policy, in regard to spectrum management. Security and Privacy Controls: Enhanced defense against the adversary attacks and data recovery of AI-supported spectrum management processes [5], [10], and [16].

Regulatory Compliance: The release 18 hosts transparency and explain ability requirements within the AI models based on regulatory provisions of AI governance to create better compliance and audit.

9.3 SECURITY AND PRIVACY, ETHICAL, AND ESG CONSIDERATIONS IN MODERN ORGANIZATIONS

9.3.1 Introduction

The RF spectrum is a limited and strategically necessary resource that provides the backbone of modern digital communications. All forms of digital communication (from mobile broadband to satellite systems to broadcasting services to communications for defense) and emerging IoT networks rely on having access to properly allocated RF spectrum. Dynamic demand for wireless communication requires that the management of RF spectrum has evolved from static, paper-based administrative processes to highly automated, data-driven digital ecosystems. While this transformation has improved the efficiency and flexibility of spectrum management, it has created various security and privacy concerns that must be addressed [5], [10], and [16].

National and international regulatory bodies (such as the FCC and ITU) regulate spectrum allocation, licensing, and interference mitigation processes. In the past, spectrum bands were allocated for long periods through exclusive licensing under rigid rules and regulations. However, today, many of the newly developed innovations rely on dynamic spectrum access (DSA) and shared licensing models, and most will require real-time coordination through digital platforms. These new types of processes are managed through spectrum management systems (SMS), which consist of a combination of cloud storage of databases (often referred to as "cloud-based"), geolocation services, automated decision-making engines, and, in some cases, AI. Although these systems can greatly enhance the agility of an organization and its operations, they also create an expanded attack surface for potential malicious actions [4], [13].

9.3.2 Security and Privacy

The recent increase in the need for security and privacy in any Spectrum Management Systems (SMS) has made them an integral part of the daily operations and innovations of many modern organizations that rely on wireless networks. As businesses begin implementing private 5G, the Internet of Things, and share-spectrum solutions, they can no longer rely on the Federal Communications Commission and International Telecommunication Union as the only entities responding to spectrum management. Many organizations are now creating their own internal systems to help ensure that spectrum resources are monitored and allocated as efficiently as possible [5], [10], and [16].

From the perspective of security, three factors to protect (i.e., confidentiality, integrity, and availability) are critical. As such, SMS databases store numerous sensitive types of information (e.g., frequency assignments, device identifiers, and network configurations). Failure to maintain a security posture designed to prevent unauthorized access means that either significantly negative effects will occur through disclosure of existing infrastructure vulnerabilities, or access to critical services will be interrupted.

When it comes to integrity, it is imperative to recognize that automated and dynamic spectrum environments rely heavily on the integrity of the data they are using; obtaining false (i.e., manipulated) sensing data, or executing a compromised allocation algorithm will most likely result in the creation of interference or unauthorized; therefore, having a robust framework including validation mechanisms, real-time monitoring of spectrum resources, and employing "zero trust" security models to minimize these risks will assist organizations in ensuring the integrity of their information.

With respect to availability, having redundancy systems and disaster recovery plans in place to support operational activities are critical to maintaining business operations. Privacy issues arise when considering that many spectrum systems collect and maintain geolocation and usage information; therefore, ensuring that organizations apply data-minimization principles, and that geolocation and usage data are maintained under secure protocols is vital to being responsible stewards of the data collected. Furthermore, having a foundation of security-by-design and privacy-by-design principles significantly enhances an organization's overall SMS system security and privacy posture.

9.3.3 Ethical Considerations

As wireless communication has become so reliant on for businesses to operate and grow, ethical considerations surrounding spectrum management and the wireless communications industry are becoming ever more important. Spectrum is a limited resource that is shared by everyone, and there are social, economic, and technological implications associated with how spectrum is allocated and used. Therefore, organizations that manage or use spectrum must take care to act ethically to promote fairness, transparency, and accountability in their actions.

A primary ethical issue is equitable access, particularly when organizations utilize their own private network(s) or participate in shared-spectrum frameworks. Organizations must avoid monopolizing specific frequencies in such a way that smaller companies or community members do not have equitable access. By using spectrum ethically, organizations will be creating an environment that fosters competition, innovation, and greater digital inclusion rather than perpetuating existing inequities in society.

Another area of concern involves transparency in decision-making. Both automated allocation systems and AI-based optimization systems should be governed by clear and predictable policies. All stakeholders must understand how spectrum management decisions are reached, especially when these decisions impact service quality, pricing, or public safety communications. Ethical practices require continuous auditing of algorithms to ensure that there is no bias or unintended discrimination involved with their use.

Another area of concern is privacy. Spectrum management systems frequently collect geolocation information, device identifiers, and usage patterns as part of their operations. Organizations must collect geolocation information, device identifiers, and usage pattern data only when it is required to operate the system; use only what is necessary; and protect that information from being misused.

9.3.4 ESG Consideration

The focus on environmental, social, and governance (ESG) factors in spectrum management is on the rise among modernized organizations. Due to digital transformations enabled through wireless connectivity, smart infrastructure, and industrial automation, organizations need to align their spectrum strategy with sustainability principles and responsible business practices.

In terms of the environment, efficient use of the spectrum will result in energy efficiency. Wireless networks operate optimally to reduce unnecessary infrastructure, lower power consumption, and allow for the deployment of smart energy systems. In addition, by deploying advanced monitoring and dynamic allocation technologies, organizations can decrease their carbon footprint associated with telecom networks and data transmission. Furthermore, sustainable procurement of telecommunications equipment and proper e-waste management will enhance environmental sustainability.

The socially oriented aspects of spectrum management include fairness related to spectrum allocation, digital inclusion for under-served communities, and community impact. Organizations need to ensure that their spectrum usage does not unfairly impact smaller organizations and under-served communities. Organizations can promote connectivity, innovation, and economic participation through policies of transparency around how they intend to allocate spectrum and through their responsible deployment of private networks. Protecting user privacy and ensuring secure communications can facilitate the establishment of public trust and a sense of social responsibility.

Good governance is integral to ethical spectrum management. A clearly defined accountability structure, compliance with regulatory requirements, and the existence of transparent decision-making processes will help achieve ethical spectrum management. Organizations should establish processes that provide strong oversight mechanisms to prevent the potential misuse of spectrum, anti-competitive activity, or regulatory non-compliance. Governance frameworks can be further strengthened through regular audits, risk assessments, and through stakeholder engagement.

9.4 SPECTRUM DATA GOVERNANCE AND SPECTRUM DATA LAKES

9.4.1 Introduction

The rapid growth of wireless technologies has turned spectrum management into a data-heavy industry. Modern wireless systems create very large amounts of technical, operational, regulatory, and geographic data. Responsible management of such data is no longer a "nice-to-have"; rather, it is an essential element of effective spectrum allocation, interference mitigation, compliance tracking, and strategic planning. As regulatory authorities and businesses implement digital spectrum management systems (SMS), the concepts of spectrum data governance and spectrum data lakes have become key drivers for transparency, efficiency and resilience[4],[13].

The data managed by organizations responsible for spectrum such as Federal Communications Commission (FCC) or International Telecommunication Union (ITU) are comprised of large numbers of files with complex datasets covering many types of information. For example, data regarding licensing, technical information, satellites, sensing data, and usage statistics can all be included. When data is not structured using established governance guidelines or stored using scalable storage systems, it can become disorganized, unreliable, or unprotected.

This chapter discusses the principles, design, advantages and difficulties of spectrum data governance and spectrum data lakes within current spectrum management framework.

9.4.2 Understanding Spectrum Data Governance

Spectrum data governance includes policies, standards, processes, and accountability mechanisms to ensure that all data related to the spectrum is accurate, safe, accessible, and in compliance with regulatory or legal requirements. It clarifies who owns the spectrum data, who can have access to it, how it should be maintained and for how long it should be kept. Data classification is the first step in effective governance.

Examples of data sets in the spectrum may be:

- Licensing and Authorization Records
- Frequency Allocation Tables
- Devices Registrations and Certificates
- Interference Reports
- Spectrum Sensing Measurements
- Geolocation and Propagation Model Data
- Enforcement and Compliance Documentation.

Each of these categories will have various levels of sensitivity. For example, Military or Public Safety Frequency Assignments must remain confidential; however, some allocation tables are available to the public.

Governance frameworks define the Standards of Metadata, Validation Rules, and Data Quality Controls. By establishing standardization across Agencies and Operators and Cross Border Partners, interoperability will be achieved. Data Stewardship Roles will provide clarity for the prevention of duplicating records and will reduce the risk of maintaining inconsistent records.

9.4.3 Key Principles of Spectrum Data Governance

- Data Integrity/Quality

To make reliable decisions about the use of the spectrum (the radio frequency) requires obtaining accurate and timely data. Necessary governance mechanisms include automated verification of data, audit trails, and version control to maintain trust in the allocation and coordination of spectrum use.

- Security/Access Controls

As the databases containing spectrum information may contain sensitive to brand and company, the spectrum databases must incorporate role-based access controls, encryption and infrastructure-based identity protection provisions to guard against unauthorized disclosure or modification of sensitive spectrum information.

- Transparency and Accountability

Public confidence in how spectrum is managed relies upon having access to transparent practices when dealing with managing spectrum resources. The design of governance policies should lay out the definition of which datasets will be made available to the public and the degree of restrictions placed on each type of dataset along with processes used to document each decision made regarding the use of an item of spectrum. Transparent processes minimize disputes between groups contributing to the administration of the spectrum and create an environment promoting equitable competition.

- Compliance and Legal Alignment

Spectrum data management must meet statutory obligations as defined by telecommunications regulations, cybersecurity standards, and data protection laws. Governance frameworks support the alignment of spectrum data management practices with domestic and/or internationally recognized statutory obligations [5], [10], [16].

9.4.4 The Development of Spectrum Data Lakes

As the complexity of the spectrum ecosystem increases, traditional relational databases may not be able to handle high volume, velocity, and variety of data in the same manner that a spectrum data lake can store this data. A spectrum data lake is defined to be a central repository capable of storing structured, semi-structured, and unstructured data from the spectrum ecosystem in its native format. This is fundamentally different than traditional databases in which an organization must use predefined schemas to ingest device network sensor feeds, device network spectrum monitoring logs, satellite device network telemetry, and operational and analytical outputs.

Data lakes are an important part of spectrum management due to:

- Large quantities of spectrum sensing data
- Real-time monitoring from distributed sensor networks
- Historical records about interference
- Analytical datasets for predictive analytics
- Training data for ML

Spectrum data lakes provide a single place to consolidate multiple different types of datasets related to spectrum use and therefore allow for more sophisticated analytics, development of policy models and automating decision-making.

Architecture of a Spectrum Data Lake: A typical spectrum data lake architecture consists of several layers

- *Layer 1 – Data Ingestion*
  This layer collects data from different sources like spectrum monitoring stations, IoT devices, regulatory databases, and other

scientists/engineers across the network operators' infrastructure and satellites. Data is automatically ingested into the database through ingestion pipelines that allow for continuous access to data.

- *Layer 2 – Storage*
  The storage of the data occurs within distributed (geographically/regionally located), scalable, often a cloud environment. The storage of data must also adhere to redundancy, high-availability, and data encryption.
- *Layer 3 – Metadata and Cataloging*
  Without proper indexing and tagging, data lakes can become ungovernable "data swamps." Therefore, metadata, indexing, tagging, and cataloging are required to provide a means of efficiently retrieving data and for governing (monitoring) the data.
- *Layer 4 – Analytics and Processing*
  Analytics use sophisticated analytics tools to examine data for patterns of spectrum occupancy, identify interference anomaly occurrences, predict demand for spectrum, and simulate the allocation of spectrum resources. Additionally, AI methods may also be applied to the analytics processing layer.
- *Layer 5 – Governance and Security*
  As an overall shared layer across all other layers of the Data Management Architecture, the governance and security layer add corporate-wide access control, logging, auditing, compliance monitoring, and data lifecycle management functionality[5],[10],[16].
- *Benefits of Spectrum Data Lakes in Spectrum Management*
  Spectrum Data Lakes are extremely beneficial to modern spectrum management, allowing for the central storage and analysis of all types of datasets (for example, licensing records, spectrum sensing data, interference reports) in one location. They allow for the consolidation of structured and unstructured data (as opposed to being stored in multiple different locations), making it possible for organizations to have better visibility into their spectrum usage patterns and overall network performance. This provides organizations with the ability to make quicker decisions regarding the detection of interference and the estimation of future demand. Additionally, data lakes allow for the use of advanced analytics and machine learning applications, which can lead to more efficient allocation (and identification) of underutilized

bands. In summary, data lakes improve an organization's operational efficiency, increase transparency, and improve overall strategic planning, while supporting a new, dynamic data-driven approach to governance of the spectrum[1],[5],[6].

- *Governance Challenges in Spectrum Data Lakes*
  In spectrum data lakes, there are many governance issues because of how complex the data is, its size, and the sensitivity of the information it contains. Without proper oversight of the data lakes, they can become disorganized, which causes record inconsistency and ultimately results in poor-quality data. To ensure that we have high-quality data, we need to establish clear validation standards, metadata management processes, and identify stewards for the data. The centralized storage of licensing information, geolocation data, and operational metrics means that our data lakes will be attractive to cyber criminals looking for ways to exploit them. Similarly, our data has privacy-related concerns associated with the physical and digital access to the data, and limiting how much personal information we are storing is one way to help mitigate these risks. Other governance issues revolve around our regulatory compliance or cross-border data transfer requirements that arise in these very large and interconnected spectrum environments.

9.4.5 Integrating Governance and Data Lake Strategies

For proper and efficient governance of the spectrum, it is necessary to establish a robust strategy for the governance and integration of data lakes. A proper governance structure aligns with the scalability of the data lake and provides analytical capabilities that can be leveraged through established policy, accountability, and compliance frameworks. Designing and embedding governance into the data lake will allow for establishing data ownership, file stewardship roles, access control, and data lifecycle management, all while maintaining data quality and security. Automated auditing, monitoring, and policy enforcement of the data lake takes place to protect against misuse, errors, and/or noncompliance. The alignment of governance with the design of the data lake will allow organizations to take advantage of the vast amounts of spectrum data for informed decision-making while ensuring the operation of their organizations in a secure and transparent manner[5],[10],[16].

9.4.6 Future Directions

Innovative new solutions like Federated Learning, Distributed Ledger Technologies, and Secure Multi-Party Computation are likely to facilitate better governance of spectrum data. They offer a means of allowing multiple parties (e.g., Regulators, Operators, Enterprises) to work together securely without disclosing their confidential raw data to one another. At an international level, the development of standardized data models may improve cross-border collaboration and interoperability. As wireless ecosystems grow with the introduction of 5G, satellite constellations and IoT networks, the management of spectrum data will also become increasingly important [1],[2],[3].

PRACTITIONER'S TAKE AWAY

AI, ML, and blockchain technologies will work together to change how we manage the available frequency spectrum in future wireless communication systems. Current spectrum management models rely on static allocation of frequencies along with centralized control. These models cannot meet the current demand for wireless communication, especially with regard to 5G, future 6G, and large-scale IoT (Internet of Things) environments. AI and ML can provide intelligent methods of spectrum sensing, traffic prediction, interference mitigation, and dynamic spectrum allocation, which will allow networks to be adaptive to varying conditions and enable better overall utilization of the available frequency spectrum.

Utilization of these technologies would also be enhanced by implementing a decentralized, transparent, and secure method of managing the use of the available frequency spectrum among all relevant stakeholders through blockchain technology and its related methodologies (i.e., distributed ledgers and smart contracts). While there are still several challenges (including scalability, regulatory adaptation, and data protection) that must be addressed to fully realize the benefits of AI and blockchain, the potential of their combined use could create intelligent, secure, and efficient spectrum ecosystems upon which future wireless networks can be developed.

REFERENCES

[1] X. Cao, "AI-Empowered Multiple Access for 6G: A Survey of Spectrum Sensing, Protocol Designs, and Optimizations," *Proceedings of the IEEE*, vol. 112, no. 9, pp. 1264-1302, 2024, doi: 10.1109/JPROC.2024.3417332.

[2] H. F. Alhashimi et al., " Survey on AI-Enabled Resource Management for 6G Heterogeneous Networks: Recent Research, Challenges, and Future Trends," *Computers, Materials & Continua,* 2025, vol. 83, Issue 3, Pages 3585-3622.

[3] A. Patil et al., "A Comprehensive Survey on Spectrum Sharing Techniques for 5G/B5G Intelligent Wireless Networks: Opportunities , challenges and future research directions", *Computer Networks(Elsevier)*, 2024, vol. 253, p. 110697.

[4] F. R. V. Guimarães et al., "Machine Learning for Spectrum Sharing: A Survey," *Foundations and Trends® in Networking*, 2024, Vol. 14, no. 1-2, pp 1–159. DOI: 10.1561/1300000037.

[5] S. E. Abdelbaset et al., "Deep Learning-Based Spectrum Sensing for Cognitive Radio Applications," *Sensors*, 2024,vol. 24, no. 24, 7907.

[6] M. A. Aygül, et al., "Machine Learning-Based Spectrum Occupancy Prediction: A Comprehensive Survey," *Frontiers in Communications and Networks*, 2025, vol. 6, https://doi.org/10.3389/frcmn.2025.1482698.

[7] M. U. Muzaffar et al., "A Review of Spectrum Sensing in Modern Cognitive Radio Networks," *Telecommunication Systems*, 2023, vol.85, pp. 347–363.

[8] P.-M. Mutescu et al., "AI-Native PHY-Layer in 6G Orchestrated Spectrum-Aware Networks," *Sensors*, 2025, vol.25(23), 7206 https://doi.org/10.3390/s25237206.

[9] A. Alauthman et al., "Deep Reinforcement Learning-Driven Dynamic Spectrum Access in Dense Wi-Fi Environments," *IEEE Access*, 2025, vol.13, pp. 178663-178680, doi: 10.1109/ACCESS.2025.3621489.

[10] J. Ye, et al., " A Trust-Centric Privacy-Preserving Block chain for Dynamic Spectrum Management in IoT Networks", *IEEE Internet Of Things Journal*, 2022,vol. 9, no. 15, pp. 13263-13278.

[11] N. V. Huynh et al., "Generative AI for Physical-Layer Communications: A Survey," *IEEE Transactions on Cognitive Communications and Networking*, 2024, vol. 10, no. 3, pp. 706-728, June 2024, doi:10.1109/TCCN.2024.3384500.

[12] M. K. Hasan et al., "Federated Learning for enables 6G communication technology: Requirements, applications and integrated with intelligence framework," *Alexandria Engineering Journal*, 2024, vol.91, pp. 658-668.

[13] M. Falco et al., "AI-Driven Spectrum Sensing: An in-depth meta-analysis of trends, challenges and opportunities," *Computer Networks*, 2026, vol. 275, Article: 111789, doi: 10.1016/j.comnet.2025.111789.

[14] Z. Li et al., " A Multi-Agent Deep Reinforcement Learning Based Spectrum Allocation Framework for D2D Communications ," *2019*

IEEE Global Communications Conference , 2019, pp. 1-6, doi: 10.1109/GLOBECOM38437.2019.9013763.

[15] F. Ali et al., " 6th Generation Wireless Spectrum Applications, Opportunities and Challenges," *International Journal of Computer Theory and Engineering* , 2025, Vol. 17, No. 2, pp. 61–82, DOI:10.7763/IJCTE.2025.V17.1369

[16] H. Xu et al., "Block chain-enabled resource management and sharing for 6G communications," *Digital Communications and Networks*, 2020, vol. 6, Issue 3, pp. 261–269.

CHAPTER 10

Review on Leveraging Blockchain for Intelligent Traffic Systems in 5G to 6G Evolution

Komala Soares, Arundhati A. Shinde, Mangal V. Patil, and Deevyankar Agarwal

10.1 INTELLIGENT TRANSPORT SYSTEM (ITS)

The continuous progress in transportation systems can be attributed significantly to the influence of advancement in communication protocol and information technologies. Advancements in intelligent transport system (ITS) over the decades have shown significant changes in efficiency in transportation, safe and sustainable network of connected vehicles, with instantaneous data exchange and deriving suitable solutions. Mohamed et al has provided insight initiatives undertaken by industries to follow pre-set standards of sustainability and safety [1]. In broad sense, an ITS is a conceptual model that encompasses network of sensors, data acquisition and processing, communication networks with the aim of bringing improvements in traffic management in smart city networks. The main objectives lie in progress and evolution in navigation system, accident prevention mechanisms, toll collection methods, traffic congestion avoidance, vehicle tracking and tracing, and many others.

 DOI: 10.1201/9781003731597-10

ITS with 3G and 4G, were limited by high latency, lower throughput, higher response time. However, the sensors like GPS (global positioning system), LiDAR (light detection and ranging), camera, and RADAR (radio detection and ranging) could not sustain 3G and 4G networks due to data rate constraint. These sensors needed improved communication protocol, which paved path for inclusion of 5G in ITS applications. Also, the incorporation of fifth-generation communication into ITS has transformed the road map by facilitating better improvements in reliability, bandwidth usage, latency, and large-scale connectivity in smart city environment. ITS and 5G have reinforced mutually, churning new potential opportunities towards autonomous driving, dynamic vehicular management, predictive analysis, and vehicle-to-infrastructure (V2X) communication stepping towards the aim of intelligent mobility of traffic [2].

10.1.1 Role of Pre-5G Networks in ITS

Prior to launching of 5G, ITS model was operating in 3G/4G communication standards, with limited sensors and centralized managing/processing centres to process and analyse the sensor data. The system was confined to finite applications like traffic monitoring and navigation support with several limitations as follows:

- High Latency: The delay in communication was in terms of 10ms to 100ms, which limited process of decision-making, which is crucial for collision avoidance.
- Centralized Architecture: Pre-5G networks operated on centralized servers, which is prone to failure of entire system from server itself. Hence, reliability and system stability were in question.
- Limited Bandwidth: The sophisticated sensors like light detection and ranging, RADAR, LiDAR, and camera collect huge amount of data, essential in decision-making. Limited data capacity of 4G networks proved not suitable for data-rich sensors. Also, vehicle-to-vehicle communication was impractical.
- Heterogeneous Sensor Integration: Heterogeneous sensors collect enormous real time data and faced challenges while storing data for further processing.

These limitations accentuated the requirement of better and suitable communication standard, which is capable and would mitigate the challenges faced by 4G networks.

10.1.2 5G-Driven Paradigm Shift in ITS

The ITS applications turned dramatically into decentralized, data-rich systems with integration of 5G communication protocol. The reduced latency of near 1ms, made it possible to achieve V2V and V2I communication with many more value-added features [3]. Large-scale connection of vehicle gave an edge to 5G-enabled ITS. Millions of vehicles within one square kilometre could be connected to the system without glitches, accomplishing high scalability, supporting connection of vehicles, roadside units (RSUs), and data-rich sensors. Popular applications like smart intersections, real-time traffic handling, re-routing suggestions, and congestion control were implemented, by employing distributed sensor networks. 10 Gbps data throughput is one of the characteristics of 5G communication protocol, which was a necessary criterion for high-end sensors like camera, LiDAR, and RADAR. These data-rich sensors were effortlessly interfaced to system with 5G, enabling advanced driver assistance systems (ADAS) and autonomous navigation. Over the years, the ITS with 5G has proved reliable ensuring efficient communication with traffic of smart city environments, with minimal interruptions. The concept of network slicing is important when network encompasses multiple, independent sub-networks with each one is customized for special task. This concept suitable in ITS, as it has various category of applications like, priority communication, emergency issues, two-way information conveyance, and entertainment applications. Network slicing can be executed successfully in ITS with introduction of 5G.

10.1.3 Suitability of 5G for ITS system

The incorporation of 5G technology with ITSs proved to be boon in providing better driving experience in smart city ecosystems. A foremost benefit is strengthened road safety, with highly reliable, lower-latency vehicle-to-infrastructure communication. Vehicles were empowered for proficient exchange of emergency alerts and notifications within milliseconds, with faster real-time responses. This instant and prompt data transfer reduced risk of accidents and improved safety.

Another major benefit is improvement in traffic efficiency and reduction in traffic congestion. ITS enabled with 5G takes advantage of AI algorithms and predictive analytics to implement many services like, dynamic traffic management, rerouting, so on. Data streams from heterogeneous sensors, from network of connected vehicles, RSUs, and other relevant infrastructure is utilized to train suitable Machine Learning models to enable pre-emptive measures. Subsequently, the congestion is reduced, time to reach is improved, start-stops are reduced, fuel consumptions are improved, carbon pollutants decreased, contributing towards better and eco-friendly environment.

Autonomous driving is mainly upheld by 5G infrastructure. Low-latency, high-speed communication enables autonomous vehicles to recognize and monitor their surroundings and take decisions within near millisecond. Cooperating vehicles of the network exchange acquired information collectively with peer vehicles for smooth and synchronized flow. Further minimization in latency is achieved through MEC (multi-access edge computing) technique, which benefits in decision-making for accident prevention, traffic coordination, and autonomous control. 5G-based ITS enhances energy efficiency from an environmental aspect, by refining routing, minimizing start-stop times, and manoeuvring traffic flow efficiently, thereby lessening carbon foot print and wastage of fuel.

Moreover, driving experience and scalability are improved. Increased data throughput maintains uninterrupted infotainment, and custom-made services. Network slicing with 5G architecture equip ITS networks with scalability and to meet any special and specific demands. With these capabilities, ITS with anticipated assimilation of 6G networks, further strengthened towards traffic safety, system integrity, efficiency, scalability, sustainability, and improved pedestrian satisfaction.

10.1.4 Limitations and Challenges of 5G for ITS System

The implementation of 5G-configured ITSs transit towards improved intelligent transport network. In spite of its potential, widespread utilization grapples with notable challenges. Significant among them is huge cost of infrastructure. 5G depends on small-cell networks, which are dense, fibre-based, and with sophisticated base stations, high-cost installation, regular maintenance, and high energy requirements. Cumulative cost makes the installation and commissioning in low-income areas economically challenging. This may be circumvented with support of government or public/private organizations.

Interoperability raises yet another important issue of coexistence of new systems with earlier installed system. This requires formation of standardized protocols and consolidated architectures. ITS standardization at global level is limited, potentially limiting cross-border automobile communication. Formulating efficient models to bridge heterogeneous communication protocols is at most requirement to enable scalability of ITS deployment.

Privacy and security of ITS remain vulnerable which requires attention. Network of vehicles with big, sensitive data generated by the infrastructure, including personal driving location and patterns, are prone to cyberattacks. Security takes a prime importance. Providing end-to-end security needs encryption, multi-level authentication, and real-time threat detection. Without these protections, passenger safety or traffic operations may be interrupted and disturbed.

Optimization of performance and efficient utilization of bandwidth pose further challenges. 5G protocol performs at millimetre-wave offering high bandwidth. Yet, it suffers from restricted range and weakening by obstacles. In case of networks of fast-moving vehicles, frequent cell transitions look for very low latency for efficient performance with respect to mobile strategies and handshaking algorithms [4].

Despite limitations, there are ongoing pilot initiatives at global level, substantiating 5G ITS potential. Significant among them the 5G Automotive Association (5GAA) at European region confirmed use of 5G for low-latency V2X applications utilized for adaptive cruise control [5]. 5G smart roads at China incorporated edge computing technology for autonomous shuttles with continuous control and monitoring, while smart city projects at U.S. upgraded traffic signals and avoided congestion through 5G-empowered roadside units [6, 7]. These works confirm that 5G suitable for existing ITS to make them capable for new applications which were out of reach of pre-5G networks. Core features of 5G facilitate mission-critical as well as data-intensive applications. Enhanced mobile broadband one of the important applications of 5G network, enabling infotainment and efficient traffic surveillance. Ultra-reliable low latency makes sure to provide critical application with respect to safety. Massive machine-type communication service brings millions of internet-connected IoT sensors together for massive automobile monitoring.

Looking further for advancements in ITS, 6th generation of wireless communication technology combined with artificial intelligence, blockchain technology, edge computing, and concept of digital twins assures latency at microsecond, throughput of terabyte and privacy secured management of traffic. This improved technology will bridge gap between physical and digital systems, providing autonomous and sustainable fleets, with holographic communication and intelligent movement of vehicles.

10.1.5 Scope of 5G While Paving for New Technology

The blending of 5G technology with ITSs represents a fundamental step towards accomplishing autonomous, smart, secured, and sustainable system. But, challenges like infrastructure costs, extent of data security, and process of standardization demands consideration to utilize its full potential. As 5G technology matures and aggregates with 6G technology along with AI and blockchain technologies, future ITS will not leave any possibilities in achieving near-no accidents, near-zero congestion, and no emissions, marking a paradigm transformation in ITS.

While 5G suffers from limitations and challenges, the paradigm shift from 5G to 6G, to overcome existing challenges is unavoidable. 6G will make sub-millisecond latency a reality, terabit-per-second a nominal speed, and AI-based decision-making. The use of blockchain will fortify the confidence in the network, keeping intruders at bay. With 6G simulation of digital twins will be lot easier for pro-predictive analysis for faster implementation. 6G will be capable of incorporating Federated learning techniques in ML/AI models for preserving privacy of individual data. With such 6G enhancements, real-time 3D mapping and holographic communication are possible, for improvements in safety and efficiency to remarkable levels.

10.2 EVOLUTION OF ITS WITH 6G

Since the onset of concept, the ITS has undergone tremendous advancements along with evolution in communication technologies. Communication plays as an integral, inseparable part in evolving ITS. With 3G communication standard, ITS was at its basic functionalities, while 4G strategies through LTE contributed better data throughput and traffic monitoring established real-time. Subsequently, 5G standard harnessed exceptionally minimal latency communication in Vehicular communication. Regardless of benefits, 5G-enabled ITS facing communication challenges.

Ever growing on-road automobiles, slow developing road structure, large storage demanding sensors and big-data-based services left no option to explore advancements to follow 5G communication standard. 6G (sixth generation) communication protocol is expected to mitigate limitations posed by 5G in ITS, with ever higher data rates, less than millisecond latency, offers further innovative and revolutionizing potential of ITS and its services.

10.2.1 Cellular Networks in Transportation

The advances in ITS have been interwoven with the innovation in cellular networks strategies:

- 3G Network standards: Takes the credit in achieving basic GPS navigation, limited, and slow V2V communication, vehicle tracking, etc. Yet, limitations of latency and bandwidth made 3G standards inefficient for real-time applications. Scalability of the system was a question.
- 4G LTE standards: However, 4G supported better data transmission speed, real-time infotainment was lagging with low speed. Cloud-based applications require high data rate transmission. This posed constraints on implementation of large-scale automobile vehicle network.
- 5G Network protocol: 5G established ultra-reliable, extremely low latency, and large-scale mass communication with enhanced bandwidth, paving many services like HD sensors installation and usage, predictive traffic control and management. Vehicle tracking and automation in driving.

Despite advanced protocol, 5G failed in applications involving large-scale terabit data, implementation of holography in communication, high-speed mobile vehicles and adoption of advanced AI algorithms for decision-making. These limitations lay foundation for incorporation of 6G [8].

10.2.2 6G: Features and Technological Innovations

The sixth generation of communication protocol, outperformed 5G in reliability, speed, and latency. Precisely, 6G excelled in addressing the challenges faced by 5G-enabled ITS. Following are prominent features of 6G standard, made it suitable communication standard for ITS.

- High data rate with terabit-per-second
- Transmission delay is less than a millisecond
- 6G standard is competent for AI processing and implementation
- Proficient in holographic and physical communication
- Security can be beefed up by employing techniques like blockchain and cryptography

In general, 6G communication protocol reinvents vehicular network for smart city application, with intelligent automatic transportation system.

10.2.3 Contribution of 6G Towards ITS Advancements

The incorporation of 6G protocol with ITS marks an advent of transformed smart city with commendable and impressive applications and services. Real-time interaction, sub-millisecond responses, high-speed communication, AI models for decision making gains appreciation. 6G contributes towards better implementation of smart grids, facilitating communication between consumers and producers, towards load balancing and energy integration. High-definition, data-rich sensors bring in efficiency to smart waste management, with AI-driven optimization in every step of waste management towards sustainability planning. By and large, 6G functions as the intelligent and integral part in smart cities promising driving efficiency, robustness and sustainability across every aspect of smart city applications and transportation system [9, 10].

10.2.4 Challenges and Limitations

Despite high hopes of innovation and improvements, the coupling of ITSs with 6G technology poses several unavoidable limitations and challenges that must be resolved prior to deployment in large scale. The major concern is foreseen complexity of infrastructure requirement, as 6G demands dense installation of dense terahertz (THz) cells and edge computing technology, escalating the cost significantly. Coexistence of 6G with previous networks and ability of interoperability also one of the major challenge and reason for concern. This calls for standardization of network protocol, which are accepted and followed by every manufacturer and customer. The issue of energy consumption takes toll on applications and promising features of 6G protocol, rendering it crippled with energy consuming transmission at THz range and AI models. The never-ending debates and discussion on legalities of data ownership, managing the allocated spectrum, liability, and responsibility of errors in autonomous vehicle contribute to slow pace and ambiguity of integration of ITS with 6G [11].

10.2.5 Future Advancement of ITS with 6G

The progression of sixth-generation (6G) communication will bring gradual transformation and development into existing ITSs. Forthcoming ITS will be identified by effortless integration decentralized intelligence intelligent real-time data analytics and seamless connected network of vehicles. One of the most cutting-edge progresses is the advancement in digital twins' simulation in real-time for traffic systems applications. The process found beneficial towards enabling faster modelling for smart city. By reflecting attributes of physical traffic in the virtual domain, researchers succeed in optimizing traffic flow dynamically, take prior decision on anticipated congestion. Federated learning allows AI models to learn distributively. Also, the horizontal federated learning technique, opens numerous possibilities of securing data privacy with minimum usage of bandwidth. With holographic communication and blockchain in the background, interaction between the vehicles for remote assistance, suggestion on navigation, temper-proof data exchange, toll management turn experiences into safer and intuitive [12].

10.3 BLOCKCHAIN AND ITS ROLE IN ITS

Blockchain was launched as an innovative technology for securing cryptocurrencies, with distributed records of transactions executed by authenticated member nodes, across multiples nodes. Decentralized method of record keeping was appreciated for its transparency and temper-resistant nature. The concept of decentralization, implemented authenticating algorithms, and visibility among participating nodes made it well-suitable protocol in ITS, where it involves private and vulnerable data exchange among vehicles and transportation infrastructure through cloud. The ITS and data are prone to internal as well as external treat and cyber-attacks. Blockchain plays a significant role in the framework of ITS towards advancement in trust management, integrity of private data, verification, and authentication of nodes and coordination between decentralized nodes or users [13].

10.3.1 Overview of Blockchain Technology

Blockchain is basically a distributed database managed collectively by a cluster of nodes that are responsible for validation and recording every transaction performed, using special cryptographic algorithm, agreed upon by every node, which are computation based, stake-based and are error tolerant. Each node of blockchain entails an irreversible timestamp, cryptographic checksum and record of previous transaction, guaranteeing traceability.

Blockchain mitigates the requirements of mediating agents in contrast to previous, centralized systems, by proving a secure environment, where every node can verify the authenticated transactions, independently. This concept of decentralization enhances flexibility of the system and eradicated failure at single point. In addition to it, smart contracts services, such as toll payments, insurance verification, and vehicle registration, can be automated with better efficiency.

Blockchain technology provides trustworthiness in ITS, enabling authenticated and secure interaction among variety of nodes or participants of the system, like peer vehicles, roadside units (RSUs), cloud-based systems, edge computing nodes, pedestrians, segways, and internet-connected IoTs and so on. Blockchain promises integral and auditable exchange of data, which is one of necessity features for large-scale network of connected vehicles [14].

10.3.2 Role of Blockchain in ITS

- Data Management with Concept of Decentralization: In traditional ITSs huge amount of sensor data generated was stored on centralized servers, which was prone to cyber-attacks and single point failure. Blockchain, in contrast, offers tamper-proof, decentralized, method of managing data assuring security, and reliability with collaborative verification [15].
- V2X with Enhanced Security: Protocol employed for communication between vehicle and every other entity of ITS, safeguards data exchange but also experiences cyber-threat. Blockchain improvise security through decentralizing topology [16].
- Management of Authenticated Nodes: In vehicular networks, blockchain maintains secure identity management with the concept of cryptographic keys and special addresses. Unique concept of Ring signatures enhance privacy, accomplishing transaction traceability authentication and efficiently [17].
- Management of Traffic and Smart Contracts: Self-executing applications of Blockchain can automate various ITS operations.

Many blockchain-based smart contracts developed and employed to reduce human interaction, thereby fastening the automation, efficiently [18].

- Integrity of Data and Its Auditability: Significance of blockchain is its feature of having an irreversible auditing capability. Decision-making in ITS environment solely depends on accuracy and authenticity of data, which can be achieved by incorporating Blockchain technology [19].
- Resource Sharing with Prior Consent and Incentivization: Blockchain also supports digital token-enabled incentive methods to boost data sharing withing authenticated vehicles and supportive infrastructure. The introduction of blockchain in ITS will certainly enhance data availability for further analysis and efficient decision-making [20].

10.3.3 Limitations and Challenges

In spite of rewarding potential, implementing blockchain into ITS poses various challenges that must be resolved globally

- Scalability and Latency: Blockchain networks are not suitable for primitive networks, which struggle to perform well with reference to latency and data throughput. Blockchain technology is failure in networks with high latency.
- Storage and Energy Constraints: Storing very large data from high-end sensors and vehicles is inefficient method and costly. With this respect, blockchain turns inefficient in system, where storage is limited.
- Privacy and Data Confidentiality: transparency trait of blockchain sometimes unintentionally expose private and sensitive data may be not secure, unless full proof authentication and secure cryptography algorithms are in place.
- Interoperability with Legacy Systems: Blending blockchain into existing ITS calls for compatibility. Achieving unobstructed interoperability still remains as a significant challenge.
- Regulatory and Governance Issues: Regulatory and legal standards for blockchain-based vehicular networks still grappling with improvements. Queries regarding ownership of data, concerned compliance and liability are still at stake, without which physical deployment is a mirage.

Incorporation of blockchain with AI/ML models, 6G, edge/cloud computing, and federated learning will further improve performance of ITS. Implementation of Edge blockchain architectures can reduce latency. AI-driven algorithms can be utilized to improve scalability and accuracy of decision, whereas federated learning techniques enhance data privacy across vehicular nodes of smart city eco systems. In recent years, researches are underway towards hybrid architectures and lightweight blockchain frameworks, combining goodness of off-chain and on-chain storage for efficient energy management.

The prospects of blockchain in ITS remain in striking a balance between efficiency, decentralization, privacy, and enabling trustworthy transportation system.

10.3.4 Conclusion

Blockchain is a technology filled with utmost potential to revolutionize ITSs, by dealing with long-waiting challenges, like security, integrity, trust and data integrity. Despite low latency, challenges in scalability and regulatory formulation, unending benchmarks in blockchain models are creating a path for robust ITS architectures. As the transportation sector leap forward, blockchain will contribute in realizing better and broader vision of smart city networks.

10.4 BLOCKCHAIN FOR ITS IN 6G ENVIRONMENT

The growth of ITSs has been catalysed by timely advancements in autonomous vehicle and wireless communication protocol along with improvements in sensor networks. 5G networks have promoted real-time vehicle-to-infrastructure communication. meanwhile, the emergence of sixth-generation networks encouraged advancements with highest bit rate and sub-microlevel data rates, virtual communication, advent of digital twin, and so on [21]. But, the adoption of 6G in ITS brings in new set of regulatory, operational, and technical challenges.

10.4.1 Technical Challenges of 6G

- Ultra-High Bandwidth and THz Communication: 6G networks take advantage of terahertz (THz) frequency bands to maximize data rates. Conversely, THz communication has to endure signal attenuation, limited range of propagation, and high sensitivity to fog and rain [22].

- Latency and Reliability in High-Speed Mobility: While 6G targets for ultra sub-millisecond latency, it faces challenge of inconsistent performance in ultra-speed vehicular networks with communication interruptions, latency spikes, loss of data packet [23].
- Edge Computing and Network Orchestration: 6G-enabled ITS counts on edge computing to analyse massive data collected by sensors, closer to source to decrease latency towards real-time decision-making. The challenge is about deploying edge nodes with considerable computational capacity, with sustainable energy consumption. Moreover, coordinating AI-model across heterogeneous network of vehicles with edge servers poses significant software level challenges [24].
- Scalability and Massive Connectivity: ITS is going to scale to millions of connected automobiles, sensors, RSUs, and IoT devices. Managing dense physical layer without network congestion and communication glitches is technical hurdle [25].
- Data Security and Privacy: The enormous amount big-data transmitted involved in 6G ITS networks are susceptible for risks of cyberattacks, breaches, and violations. Even though blockchain is equipped with cryptography, implementation requires careful design [26].

10.4.2 Operational and Implementational Challenges

Establishment of 6G-enabled ITS presents several infrastructural and operational constraints that must be dealt with to maintain sustainability and scalability. Infrastructure construction is one of the great challenges, as it requires installation of small cells densely populated, with terahertz (THz) antennas, edge nodes, and RSUs, supporting high-speed communication with least possible delay. Large-scale installation incurs huge cost. City planning and spectrum licensing are added complications to deployment. Also, backward compatibility to avoid network fragmentation is necessity.

Another important issue is energy efficiency in the system. As THz communication demands higher power, developing energy-minimizing communication protocols and supporting hardware becomes essential for sustainable system.

Finally, regulatory and standardization of protocol remains pivotal, demanding globally accepted policies on allocation of spectrum, data governance, and communication standards, for efficient deployment of 6G ITS.

10.4.3 Implementation and Development

- Advanced Physical Layer Technologies: Future research should focus on robust THz channel modelling, beamforming techniques, and adaptive antenna systems to mitigate signal loss and interference. Development of multi-band hybrid communication systems integrating millimetre wave, sub-THz, and visible light communication can ensure reliable coverage across diverse urban and highway scenarios [27].
- AI-Driven Network Management: AI/ML algorithms can be adopted towards optimizing resource distribution and to anticipate vehicular mobility, detect anomalies, and enhance network efficiency. Research on federated learning, reinforcement learning, and edge AI is essential for developing self-organizing ITS networks capable of real-time decision-making [28].
- Edge-Cloud Integration and Digital Twins: Integrating edge computing with cloud infrastructure allows low-latency local processing while leveraging global intelligence for predictive analytics. Creating digital twins of transportation systems can simulate traffic scenarios, optimize fleet management, and anticipate congestion, thereby improving operational efficiency [29].
- Secure and Privacy-Preserving Protocols: Further work is needed to design lightweight encryption, blockchain-based identity management, and homomorphic encryption for 6G-enabled ITS. These protocols must balance security and privacy with real-time processing constraints [26].
- Sustainable Deployment Strategies: Researchers should work on optimizing the energy consumption so as to reduce ill effect on environment [30].
- Holographic and Tactile ITS Applications: 6G is credited for vehicle tracking, holographic navigation, remote vehicle management. Exploration of tele-maintenance and improved ADAS should be carried out [31].
- Implications for Autonomous Vehicles: There is need for Autonomous networks to overcome challenges of security, reliability, and interoperability, along with rigorous testing and simulation [32].

- Smart City Integration: Smart city integrates in numerous applications and services. Real-time traffic management, smart parking, adaptive lighting, and energy-efficient public transport besides network of vehicles. Combining ITS network with IoT, digital twins, found to be beneficial [33].

10.4.4 Conclusion

6G communication protocol proves to be a significant advancement in ITS, assuring ultra-low latency, widespread AI integration, unmatchable data rates and massive connectivity. While 6G can ensure development of fully autonomous networks of automobiles, several challenges are not yet solved. Near future must address and mitigate these challenges through advanced designs, secure protocols, AI-based edge /cloud integration, and innovative infrastructure strategies. The convergence of 6G with autonomous vehicles, within smart city technologies has capability to revolutionize next-generation ultra-ITSs.

10.5 CASE STUDY: 5G-ENABLED, IOT-BASED OBUS FOR INTELLIGENT TRAFFIC CONTROL IN A SMART CITIES

10.5.1 Introduction

Ever growing population with fast urbanization and increased travel time on existing road network leading to traffic congestion and safety issues. This reveals that the transportation system requires advancements. The intelligent traffic management tries to solve these challenges using real-time raw sensor data, AI-driven automation algorithm, leveraging advanced communication strategies, towards smooth, safe, improvised traffic flow with minimal environmental impact. Incorporating 6G communication standards with Blockchain technology for the network of connected automobiles, ITS shows improvement to a greater extent [34].

On-board units (OBUs) provide communicational gateway for vehicle-to-everything (V2X) interactions, supporting traffic control dynamically. OBUs take part in adaptive congestion management and predictive signalling.

This case study contributes towards validating the design and implementation of IoT-based OBUs within a smart city system, reinforcing real-time monitoring, wireless communication and server-based analytics, by employing AI models. OBUs were deployed for an on-field experimentation. Data collected at server through wireless communication was preprocessed and analysed before utilizing to train AI models. Various ML/AI models were trained for classification of traffic conditions.

The results were very satisfactory. Among various ML models, Decision Tree was more accurate for the said application.

10.5.2 Proposed System Architecture

IoT-based OBU system is developed to incorporate on-the-fly communication among connected vehicles with a centralized server. Peer-to-peer communication as per Industry 4.0 standard was also established. The innovative system suggested is depicted at Figure 10.1

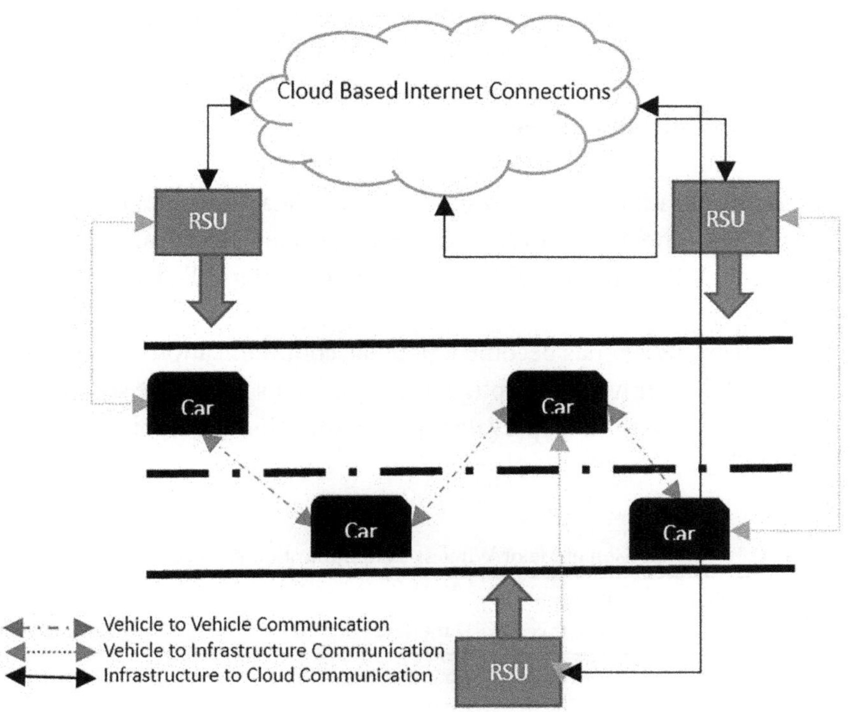

FIGURE 10.1 Overall architecture of the proposed system for moving vehicles.

The architecture is structured into basic layers: physical layer, second communication layer, and application layer.

- **Physical layer**

The physical layer comprises sensors and data acquisition modules installed on vehicles and traffic infrastructure. In the vehicle, OBUs are equipped with:

- GPS modules for real-time location tracking.
- Three-axes Accelerometer for sensing the speed in all three axes
- Connectivity modules (in-built Wi-Fi) for data transmission.

Mobile handsets are considered as RSUs with un-obstructed connection to OBUs. This arrangement not only provides gateway between OBUs and Cloud server, but also simplifies the implementation, so that intense experimentation can be done on OBU and its performance. These devices continuously monitor vehicle positions, and speeds generating multi-source real-time data streams that are fed to the communication layer.

- **Communication Layer**

The communication layer is responsible for data transmission between OBUs, RSUs, and the central traffic management platform. The system implements V2I as well as V2V communication using IoT-enabled wireless protocols.

Over years, Wi-Fi has become a popular communication protocol, due to its scalability, universal adoption and higher data rates. The summary of comparison of Wi-Fi with prevalent communication protocol are given in Table 10.1.

TABLE 10.1 Comparison of Major Wireless Communication Protocol

Communication Protocol	Range of Distance	Data Rate	Power Consumption
Wi-Fi (IEEE 802.11)	50–100 m	Up to 9.6 Gbps (Wi-Fi 6)	Medium–High
Bluetooth/BLE	10–100 m	Up to two megabits per second	less
Zigbee protocol	10–100 m	200 kilobits per second	Extremely low
LoRa/LoRaWAN	Up to 10km	less than fifty kilobits per second	Extremely low

A comparative study of various communication technologies indicates that Wi-Fi is the most relevant solution for communication between vehicles and other infrastructure of the system. Wi-Fi offers seamless integration, cost-effectiveness, high data rates, and broad device compatibility. Moreover, Wi-Fi supports mesh connectivity applicable for V2V communication.

- **Application Layer**

The application layer includes cloud platform for centralized traffic management, which provides data preprocessing, data analysis, validating the ML/AI model for decision-making, and generating visualization services. A high-performance laptop serves as the system's central server. Key components include:

- **Traffic flow analysis module:** Uses real-time data from OBUs to detect congestion and traffic patterns.
- **Signal control module:** Dynamically predicts the master control messages to all the OBUs. Typically, "Go", "Stop", "Divert".
- **Dashboard and visualization:** The master control messages are sent back to every node/OBU. OBU in-turn displays by lighting tri-colour LEDs on the surface.
- **Alert and notification system:** The notifications suggest drivers via OBUs regarding congestion, accidents, or route suggestions.

10.5.3 Prototype Design: On-Board Unit

The prototype system was implemented using commercially available microcontrollers and IoT modules. Each OBU integrates an ESP32 microcontroller, GPS module, accelerometer, and in-built wireless communication unit. Traffic management software was developed in Python with a LED display to visualize traffic data and control signals. The prototype was tested in a urban roads with five vehicle nodes as shown in Figure 10.2.

The OBU serves as the core vehicle interface for ITS, responsible for data collection, processing, and communication. The hardware design emphasizes less weight, small size, low power consumption, and modularity, allowing easy installation in diverse vehicle types.

FIGURE 10.2 Field Installation of OBU module in the vehicle and route of field experiment.

The OBU software stack is designed to perform:

- **Data acquisition and preprocessing:** Collects GPS, and Accelerometer sensor data at fixed intervals. Aggregates the data prior to preprocessing and training ML models for classification of traffic conditions.
- **Messaging protocol implementation:** various models are trained, tested for best accuracy. The validated model is deployed for real time classification of Traffic conditions. Subsequently, OBU receives traffic control commands (e.g., "Go", "Stop", "Divert") from the traffic management platform prompting the driver for better and secure driving experience.
- **V2V communication**: Wi-Fi mesh protocol is used for inter-vehicular communication. Each OBU tracks number of OBUs in close vicinity. More the number of vehicles, denser is the traffic. This situation is signalled to driver with changing brightness on Pulse Width Modulation (PWM) driven LED strip, making driver aware of surrounding.

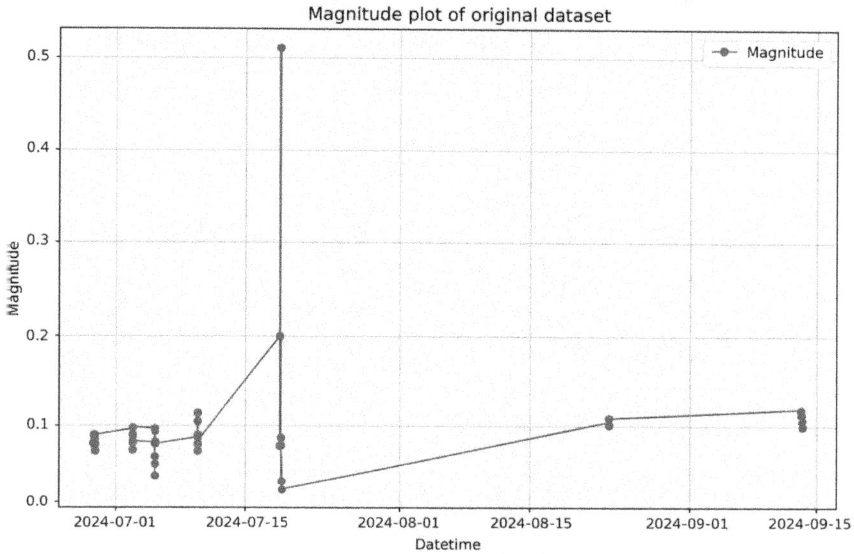

FIGURE 10.3 Magnitude plot of original data with Time stamp.

10.5.4 Working Methodology and Result Analysis

The working methodology of the IoT-enabled OBU system with 5G Protocol follows a **data-driven, real-time communication model**:

- **Data Collection:** OBUs continuously collect vehicle speed and location. The Magnitude of speed plotted against time stamp at Figure 10.3.
- **Data Processing:** Resampling ensures uniform temporal resolution by adjusting data to consistent frequencies, while polynomial interpolation of order 2, estimates missing values, maintaining continuity and synchronizing measurements across multiple OBUs or sensors as shown in Figure 10.4.
- **Data Transmission (V2I):** OBUs transmit this data through RSUs to the central server using V2I communication. server pre-process the data to generate immediate control actions for traffic signals.
- **Real-Time Analysis:** The central traffic management platform aggregates data from multiple sources, applies machine learning models for congestion prediction, and generates Master Control signals to all OBUs.
- **Models**: Various Machine Learning models are tested for accuracy. Table 5.2 summarizes accuracy, precision, recall and F1 score for

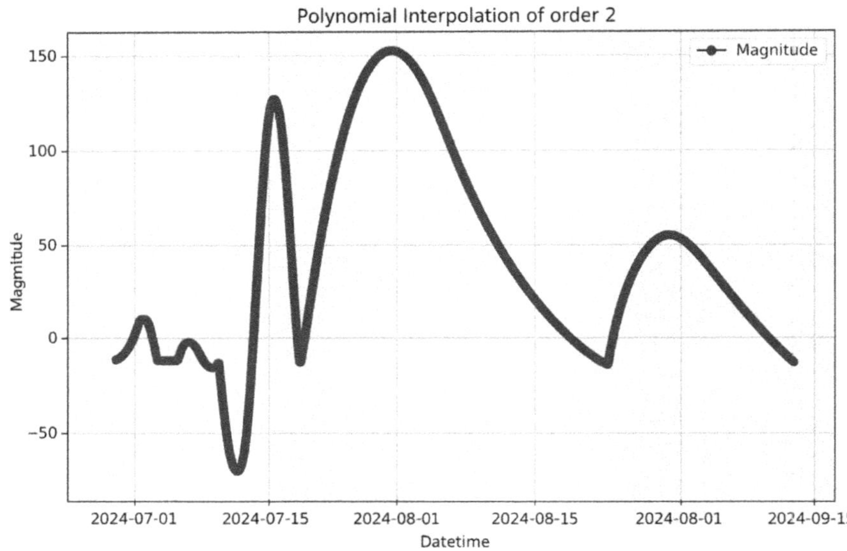

FIGURE 10.4 Magnitude plot of data after polynomial interpolation of order 2.

TABLE 10.2 Performance Metrics of Machine Learning Models

Machine Learning Model	Accuracy	Precision	Recall	F1 Score
Linear Regression	88%	89%	88%	88%
Decision Tree	88%	91%	88%	87%
Support Vector Machine	88%	89%	88%	89%
K-Nearest Neighbours	85%	87%	85%	80%
Neive Bayes	81%	83%	81%	80%
Convolution Neural Network	79%	99%	78%	77%

every model. But, Decision Tree model predicts the traffic conditions more accurately.

- **Control Command Generation:** Based on the analysis, the system generates dynamic master control signal for adjustments and rerouting recommendations
- **Command Execution:** Commands are transmitted back to OBUs. Vehicles receive real-time alerts, while traffic signals adapt automatically to optimize flow.
- **Monitoring and Feedback:** System performance is continuously monitored, and data is stored for historical analysis, performance evaluation, and system optimization.

- **Data Transmission:** Each OBU transmits its ID over Wi Fi Mesh network, in master mode. While OBUs in the range receive IDs of OBUs in slave mode. Every interval of time, OBU can sense how many OBUs/Vehicles are in range and can estimate vehicular density.

10.5.5 Conclusion

This case study demonstrates the feasibility and effectiveness of IoT-enabled OBUs for intelligent traffic control within smart city with 5G environments. The proposed system utilizes OBUs for edge computing during V2V communication. Also, IoT communication protocols and real-time analytics were used to enhance traffic control and management. The proposed prototype successfully establishes V2V and V2I communication at speed of 1Hz. The communication is found substantially reliable with adaptive traffic control in real-time traffic condition.

10.5.6 Future Scope

Future enhancements may include:

- **Incorporation of ITS with 5G/6G networks** for very low transmission delay and faster data exchange.
- **Implementation of federated Learning Protocols** for preserving data privacy.
- **Deployment of protocols** for data security and Vehicle Authentication
- **Enhanced OBUs with advanced sensors** such as LiDAR and cameras for autonomous vehicle compatibility.
- **Blockchain-based security** for tamper-proof communication and data integrity across the network.

The combination of various advanced technologies like IoT, federated learning, edge computing, authentication, and advanced communication protocol leads path for safer, sustainable, and smarter vehicular network for smart cities.

REFERENCES

[1] M. Elassy, M. Al-Hattab, M. Takruri, and S. Badawi, "Intelligent transportation systems for sustainable smart cities," *Transp. Eng.*, vol. 16, p. 100252, Jun. 2024.

[2] M. C. Lucas-Estañ, B. Coll-Perales, T. Shimizu, J. Gozalvez, T. Higuchi, S. Avedisov, O. Altintas, and M. Sepulcre, "An analytical latency model and evaluation of the capacity of 5G NR to support V2X services using V2N2V communications," arXiv preprint, 2022.

[3] E. H. Moussa, A. Amrouche, and S. T. Kebir, "4G/5G in smart cities: Enhanced analysis for intelligent transportation systems (ITS)," in *Proc. 2024 Int. Conf. on Telecommunications and Intelligent Systems (ICTIS)*, Djelfa, Algeria, 2024, pp. 1–6, doi: 10.1109/ICTIS62692.2024.10893920.

[4] L. Guevara and F. Auat Cheein, "The role of 5G technologies: Challenges in smart cities and intelligent transportation systems," *Sustainability*, vol. 12, no. 16, p. 6469, Aug. 2020.

[5] J. M. Pereira, "5G for connected and automated mobility (CAM) in Europe: Targeting cross-border corridors," *IEEE Network*, vol. 35, no. 3, pp. 6–9, Jun. 2021.

[6] Z. Fu, "The application of 5G communication technology in intelligent transportation systems," *Int. Core J. Eng.*, vol. 11, no. 10, pp. 19–24, Oct. 2025.

[7] C. H. Wang, E. Steinfeld, J. L. Maisel, and B. Kang, "Is your smart city inclusive? Evaluating proposals from the U.S. Department of Transportation's Smart City Challenge," *Sustain. Cities Soc.*, vol. 74, p. 103148, Nov. 2021.

[8] H. Karjaluoto, "An investigation of third generation (3G) mobile technologies and services," *Contemp. Manag. Res.*, vol. 2, no. 2, pp. 91–98, 2006.

[9] S. Sharma, Popli R, Singh S, Chhabra G, Saini G S, Singh M, Sandhu A, Sharma A, Kumar R, "The role of 6G technologies in advancing smart city applications: Opportunities and challenges," *Sustainability*, vol. 16, no. 16, p. 7039, Aug. 2024.

[10] M. M. Azari, S. Solanki, S. Chatzinotas, and M. Bennis, "THz-empowered UAVs in 6G: Opportunities, challenges, and trade-offs," *IEEE Commun. Mag.*, vol. 60, no. 5, pp. 24–30, May 2022.

[11] C. Han, L. Yan, and J. Yuan, "Hybrid beamforming for terahertz wireless communications: Challenges, architectures, and open problems," *IEEE Wirel. Commun.*, vol. 28, no. 4, pp. 198–204, Aug. 2021.

[12] Y. Lu, X. Huang, K. Zhang, S. Maharjan, and Y. Zhang, "Low-latency federated learning and blockchain for edge association in digital twin empowered 6G networks," *IEEE Trans. Ind. Informat.*, vol. 17, no. 7, pp. 5098–5107, Jul. 2021.

[13] A. Balasubramaniam, M. J. J. Gul, V. G. Menon, and A. Paul, "Blockchain for intelligent transport system," *IETE Tech. Rev.*, vol. 38, no. 4, pp. 438–449, 2020, doi: 10.1080/02564602.2020.1766385.

[14] K. Fan, Q. Pan, K. Zhang, Y. Bai, S. Sun, H. Li, and Y. Yang, "A secure and verifiable data sharing scheme based on blockchain in vehicular social networks," *IEEE Trans. Veh. Technol.*, vol. 69, no. 6, pp. 5826–5835, Jun. 2020.

[15] M. Firdaus, S. Rahmadika, and K. H. Rhee, "Decentralized trusted data sharing management on Internet of Vehicle edge computing (IoVEC) networks using consortium blockchain," *Sensors*, vol. 21, no. 7, p. 2410, Mar. 2021.

[16] J. Meijers *et al.*, "Blockchain for V2X: Applications and architectures," *IEEE Open J. Veh. Technol.*, vol. 3, pp. 193–209, 2022.

[17] S. Abbas *et al.*, "Blockchain-based authentication in Internet of Vehicles: A survey," *Sensors*, vol. 21, no. 23, p. 7927, Nov. 2021.

[18] A. K. Tyagi, S. Tiwari, and K. Naithani, "Blockchain, AI, and IoT for smart road traffic management system," in *Digital Twin and Blockchain for Smart Cities*, Oct. 2024, pp. 197–214.

[19] C. Chen, L. Wang, Y. Long, Y. Luo, and K. Chen, "A blockchain-based dynamic and traceable data integrity verification scheme for smart homes," *J. Syst. Archit.*, vol. 130, p. 102677, Sep. 2022.

[20] P. Liu, Z. Zhang, C. Ren, and H. You, "A blockchain-based resource sharing incentivization mechanism for multi-to-multi in compute-first networking," *Computer Networks*, vol. 266, p. 111318, May 2025.

[21] T. Hewa *et al.*, "The role of blockchain in 6G: Challenges, opportunities and research directions," in *Proc. 2nd 6G Wireless Summit (6G SUMMIT)*, Levi, Finland, 2020, pp. 1–5.

[22] A. V. Jha, B. Appasani, M. S. Khan et al., "6G for intelligent transportation systems: Standards, technologies, and challenges," *Telecommun. Syst.*, vol. 86, pp. 241–268, 2024, doi: 10.1007/s11235-024-01126-5.

[23] J. Rodríguez-Piñeiro, Z. Wei, J. Wang, C. A. Gutiérrez, and L. M. Correia, "6G-enabled vehicle-to-everything communications: Current research trends and open challenges," *IEEE Open J. Veh. Technol.*, vol. 6, pp. 2358–2391, 2025.

[24] A. M. Alwakeel, "Synergistic integration of edge computing and 6G networks for real-time IoT applications," *Mathematics*, vol. 13, no. 9, p. 1540, May 2025.

[25] A. A. Anvigh, Y. Khavan, and B. Pourghebleh, "Transforming vehicular networks: How 6G can revolutionize intelligent transportation," *J. Sci. Eng. Technol.*, vol. 4, no. 1, pp. 80–93, Apr. 2024.

[26] A. D. Abdullahi, E. Bahrami, T. Dargahi, M. Al-Khalidi, and M. Hammoudeh, "Interplay between security, privacy and trust in 6G-enabled intelligent transportation systems," arXiv preprint arXiv:2510.02487, Oct. 2025.

[27] R. Liu *et al.*, “6G-enabled advanced transportation systems,” *IEEE Trans. Intell. Transp. Syst.*, vol. 25, no. 9, pp. 10564–10580, Sept. 2024.

[28] I. Ullah *et al.*, “Autonomous network management for 6G communication: A comprehensive survey,” 11, *Digit. Commun. Netw.*, Jul. 1917–1940, 2025.

[29] M. Crespo-Aguado, R. Lozano, F. Hernandez-Gobertti, N. Molner, and D. Gomez-Barquero, “Flexible hyper-distributed IoT–Edge–Cloud platform for real-time digital twin applications on 6G-intended testbeds for logistics and industry,” *Future Internet*, vol. 16, no. 11, p. 431, Nov. 2024.

[30] J. Wang, K. Zhu, and E. Hossain, “Green Internet of Vehicles (IoV) in the 6G era: Toward sustainable vehicular communications and networking,” *IEEE Trans. Green Commun. Netw.*, vol. 6, no. 1, pp. 391–423, Mar. 2022.

[31] A. Elzanaty, F. Guerra, D. Dardari, and M.-S. Alouini, “Toward 6G holographic localization: Enabling technologies and perspectives,” *IEEE Internet Things Mag.*, vol. 6, no. 3, pp. 138–143, Sept. 2023.

[32] X. Deng *et al.*, “A review of 6G autonomous intelligent transportation systems: Mechanisms, applications and challenges,” *J. Syst. Archit.*, vol. 142, p. 102929, Sept. 2023.

[33] P. R. Singh, V. K. Singh, R. Yadav, and S. N. Chaurasia, “6G networks for artificial intelligence-enabled smart cities applications: A scoping review,” *Telematics and Informatics Reports*, vol. 9, p. 100044, Mar. 2023.

[34] A. M. Fadhil, “Next-generation urban intelligence: Integrating 6G communication and blockchain technology for smart city advancements,” *Int. J. Comput. Sci. Mobile Comput.*, vol. 12, pp. 29–35, 2023.

Index

P

Q

R

S

T

U

V

W

For Product Safety Concerns and Information please contact our EU representative GPSR@taylorandfrancis.com
Taylor & Francis Verlag GmbH, Kaufingerstraße 24, 80331 München, Germany

www.ingramcontent.com/pod-product-compliance
Ingram Content Group UK Ltd.
Pitfield, Milton Keynes, MK11 3LW, UK
UKHW021957220926
475880UK00003B/99
* 9 7 8 1 0 4 1 2 2 2 9 0 3 *